Farm Fresh Broadband

Information Policy Series

Edited by Sandra Braman

The Information Policy Series publishes research on and analysis of significant problems in the field of information policy, including decisions and practices that enable or constrain information, communication, and culture irrespective of the legal siloes in which they have traditionally been located as well as state-law-society interactions. Defining information policy as all laws, regulations, and decision-making principles that affect any form of information creation, processing, flows, and use, the series includes attention to the formal decisions, decision-making processes, and entities of government; the formal and informal decisions, decision-making processes, and entities of private and public sector agents capable of constitutive effects on the nature of society; and the cultural habits and predispositions of governmentality that support and sustain government and governance. The parametric functions of information policy at the boundaries of social, informational, and technological systems are of global importance because they provide the context for all communications, interactions, and social processes.

A complete list of the books in the Information Policy series appears at the back of this book.

Farm Fresh Broadband

The Politics of Rural Connectivity

Christopher Ali

The MIT Press
Cambridge, Massachusetts
London, England

The MIT Press would like to thank the anonymous peer reviewers who provided comments on drafts of this book. The generous work of academic experts is essential for establishing the authority and quality of our publications. We acknowledge with gratitude the contributions of these otherwise uncredited readers.

This book was set in Stone Serif and Stone Sans by Westchester Publishing Services.

Library of Congress Cataloging-in-Publication Data

Names: Ali, Christopher, author.
Title: Farm fresh broadband : the politics of rural connectivity / Christopher Ali.
Description: Cambridge, Massachusetts : The MIT Press, [2021] | Series: Information policy series | Includes bibliographical references and index.
Identifiers: LCCN 2021000474 | ISBN 9780262543064 (paperback)
Subjects: LCSH: Rural telecommunication—Government policy. | Broadband communication systems—Government policy. | Rural telecommunication— Economic aspects. | Broadband communication systems—Economic aspects. | Digital divide.
Classification: LCC HE7645 .A45 2021 | DDC 384.3/2091734—dc23
LC record available at https://lccn.loc.gov/2021000474

152941459

To Ben—my heart
To Tuna, Titan, and Stella—our home

The future belongs to the connected.
— Jessica Rosenworcel, FCC Commissioner and Acting Chair

Our future town square will be paved with broadband bricks.
— Michael J. Copps, Former FCC Commissioner and Acting Chair

Contents

Series Foreword

During the first decades of the twentieth century, the rural Midwest boasted the densest telephone penetration in the United States. The percentage of the population on the network grew rapidly after an enterprising dentist returned to Iowa from a World's Fair and set up a line linking his office to his house so he could go home for lunch. Farmers set up their own lines along rural fencing; housewives served as operators, working out of their kitchens. Seven decades later, Canada became the first government outside of the United States to enter the formal internet design process with its call to make sure that those in rural areas would have access.

AT&T was perceived as a monopoly in the early 1980s when an antitrust agreement required the company to divest itself of local telephony, but there were actually 14,000 telephone companies and cooperatives in the United States at the time, almost all rural. Although these were all separate organizations, they were experienced as a single system because nationalization of the private sector entities providing telephony during World War I made network effects—the fact that the value of a communication network increases the more people are on it—vividly clear, so interconnection was one of the first regulatory mandates put in place.

It was also in the 1980s that rural communities in a Minnesota region, grown impoverished after the mining industry collapsed, laid down their own fiber-optic cables. That was a "twofer"—the cables were first put in to enable schools to share rare resources, such as foreign language teachers. Once in, they also attracted industry and telework to the region, with significant benefit to the economy. Although it had responsibility for telecommunications, the state's public utilities commission turned its eyes away when these independently installed fiber cables were brought to its attention because, regulators said, they simply didn't know what to do about it.

When those who laid the cables were asked how they accomplished this, the answer was simple: they bought some fiber, dug ditches, and laid it. "Precision agriculture," the use of GPS to target water and fertilizer where specifically needed, developed during the same period.

Sociologists Phillip Tichenor, George Donohue, and Clarice Olien were looking at geographic as well as socioeconomic differences when they found evidence of the "knowledge gap," the relationship between access to information and socioeconomic class, in the 1970s. By the mid-1990s, this became known as the "digital divide," with a special emphasis on access to the infrastructure through which information, and knowledge, can be acquired. The digital divide exists across countries as well as within them. In this regard, the United States is no outlier. (In 2019, not only did Nordic countries such as Finland, Norway, and Sweden, come out ahead of the United States in terms of internet penetration, but so did Kuwait, Aruba, and the Falkland Islands, among others.)

It is to the geographic side of the digital divide that Christopher Ali addresses *Farm Fresh Broadband*, focusing on barriers to broadband access in the rural United States. The problem is particularly acute on tribal lands, which are excluded from Telecommunications Act of 1996 requirements for public access to the internet in schools, libraries, and medical facilities. The problem, not surprisingly, is one of money. Because of network effects, it made sense for a company with the monopoly-like reach of AT&T before the divestiture to extend service into rural areas. In a competitive environment, though, network providers can "cherry pick," serving just those from whom they earn the greatest profits and leaving the rest to "the other guys," to no one, or to the communities needing service themselves.

It is the last of these that Ali highlights in the stories he tells and in his policy recommendations. Local champions and community networking efforts of diverse kinds make a big difference, just as they did over a century ago with the telephone, but they may not be enough; Ali makes recommendations for the federal government as well. Telephones made such a difference to farmers in the early twentieth century not only for sociability—the ability of multiple households to get onto their shared party lines to tell stories and sing—but for business as well. For agriculture, as for so many industries, the telephone transformed the relationship of individual producers to the market as well as to how their production took place.

Ali explains how all of these things are true as well of broadband internet, with needs significantly intensified by the COVID-19 pandemic forcing almost all social processes online. The author refers to Steven Lukes's conceptualization of the three faces of power—instrumental (controlling the material world), structural (controlling social structures and rules), and symbolic or consensual (controlling perceptions and beliefs). Since Lukes wrote in the 1970s, we have come to recognize another face of power—informational—that involves the informational foundations of power in its instrumental, structural, and symbolic forms and offers new opportunities to exercise power digitally. All four are in play when it comes to rural broadband access.

Farm Fresh Broadband notes that the pandemic has created pressure on telecommunications policymakers to think again about their level of support for diverse populations. This book provides a wealth of nutrition for them, as for students, scholars, and the rest of us.

Sandra Braman

Acknowledgments

I always struggle with acknowledgments because I want to publicly thank everyone I have ever met or who has ever said a kind word to me. So, to all these people, I say from the bottom of my heart: thank you. With the limited space I have here, I thank the following individuals and organizations.

This book had two sources of inspiration: my previous research on local broadcasters and small market newspapers and Professor Darin Barney's (2011) article "To Hear the Whistle Blow" on the grain silo as a communication medium. I am grateful to Professor Barney for his article, his exemplary scholarship, and for inspiring me to undertake this research.

The project began to take shape in 2017 when I was a fellow at the Center for Advanced Research in Global Communication (CARGC) at the Annenberg School for Communication in Philadelphia. I am grateful to CARGC for this fellowship, to Marina Krikorian for her support, to the students and postdocs who attended the biweekly "policy and pizza" discussions, and to the incredible Dr. Marwan Kraidy, now dean and CEO of Northwestern University in Qatar, for his never-ending support of my work.

The bulk of the writing of *Farm Fresh Broadband* took place while I was a fellow at the Benton Institute for Broadband & Society in 2019 and 2020. Benton gave me an incredible platform to publish my thoughts, theories, and opinions related to all things broadband, and words cannot express my gratitude to the institute. Specifically, I thank Adrianne Furniss, Kevin Taglang, Kip Roderick, and Robbie McBeath. Benton also introduced me to amazing and thoughtful leaders and advocates in American broadband and telecommunication policy, whom today I am honored to call colleagues: Gigi Sohn, Mignon Clyburn, Andrew Schwartzman, and Joanne Hovis. Special mention is owed to Jonathan Sallet, who is a constant champion of my work and whom I consider a friend and mentor.

My research and this book would not have been possible without the people who graciously took the time to speak with me and share their experiences and wisdom. I am grateful and indebted to those I interviewed, those who invited me to events, and those who just took a few minutes to speak with me while I was conducting fieldwork. As I argue repeatedly in this book, broadband is not about technology, policy, or markets: it's about people, and I thank all those I met as part of my research.

During my fieldwork and since, I have had the opportunity to meet an array of incredible individuals and organizations, many of whom are quoted or cited in this book. I would like to especially thank Drew Clark, Doug Dawson, Christopher Mitchell, Sascha Meinrath, Harold Feld, Roberto Gallardo, Brian Whitacre, and Deb Socia for their advice and support, along with those from BroadbandNow (Tyler Cooper, Jameson Zimmer), Common Cause (Michael Copps, Yosef Getachew), Connected Nation (Tom Ferree, Jessica Denson, Ashley Hitt, Eric Frederick, David Daack), the National Digital Inclusion Alliance (Angela Siefer), Next Century Cities (Cat Blake), the National Rural Electric Cooperative Association (Paul Breakman, Brian O'Hara, Kelly Wismer, Tracy Warren), NTCA— The Rural Broadband Association (Shirley Bloomfield, Michael Romano, Laura Withers), the Schools, Health & Libraries Broadband Coalition (John Windhausen, Alicja Johnson), and the Federal Communications Commission (Commissioner Jessica Rosenworcel, Umair Javed).

My colleagues in the Department of Media Studies at the University of Virginia (UVA) are as inspirational as they are kind and generous, and I am grateful for their support. I thank all of my colleagues but would like to highlight our chair, Camilla Fojas; our office administrator, the incomparable Barbara Gibbons; and our librarian, Erin Pappas. UVA also supported the fieldwork for this research, and I am grateful to Associate Dean for the Social Sciences and Professor of History Christian McMillen, former associate dean Len Schoppa, and the Center for Global Inquiry + Innovation. Many thanks also to Parker Bach for helping me with research and to my colleagues in the Marxist reading group for keeping me inspired. My students at UVA give me life and I am fortunate to have taught a handful who share my passion for broadband. These amazing students include Lukas Pietrzak, Ben Tobin, Jordan Arnold, Cat Blake, Anna Higgins, and Katie Jordan. I am truly honored to have worked with these young people, am fiercely proud of their accomplishments, and cherish the fact that they are my friends and colleagues. Mark Duemmel was my research assistant for the most substantive years of research and writing of this book and deserves special acknowledgment. Mark started as my research assistant on the first day of his first year of college and stayed with this project for four years, until he graduated. It is not an understatement to say that I could not have done this work without him and I will forever be grateful for his help and his advice. Let this be a lesson for all of us: undergrads get it done!

Many friends and colleagues took the time to read parts of this book, including Manuel Puppis, Hilde Van den Bulck, Pawel Popiel, and Andre Cavalcante. Aswin Punathambekar, Jonathan Sallet, Victor Pickard, Mark Duemmel, and Stephen Herzenberg each assumed the daunting task of reading and commenting on the entire draft. I thank them all profusely and beg their forgiveness for asking in the first place. Work from this research was presented at multiple conferences and meetings, including various gatherings of the International Communication Association, the European Communication Research and Education Association, the Research Conference on Communications, Information and Internet Policy (TPRC), Broadband Communities Summit, and Reimagine Appalachia. I am grateful for the opportunities and feedback I received at these events. I also had the honor of presenting my work as an invited speaker at Guelph University, Drexel University, the Information Society Project at Yale Law School, the Berkman Klein Center at Harvard University, the Programme in Comparative Media Law and Policy at Oxford University, the House Democrat Task Force on Rural Broadband, and the Blue Ribbon Reimagine New York Commission. Words cannot express my gratitude for these opportunities. Early versions of chapters 2 and 3 were published as "The Politics of Good Enough" in the *International Journal of Communication* (Ali 2020) and "The Reluctant Regulator" in *Telecommunications Policy* (Ali and Duemmel 2018), respectively. I thank the editors of these journals for permission to republish.

At MIT Press, my thanks and appreciation go to Professor Sandra Braman for recruiting me to the Information Policy Series, to Gita Manaktala for shepherding my book through the publication process, Erika Barrios for being my point of contact, and to the four anonymous reviewers who provided helpful feedback. When I began to write in earnest, I partnered with Ideas on Fire, led by the brilliant Cathy Hannabach, to help conceptualize and edit my work. I could not have written *Farm Fresh Broadband* without Cathy or my editors Emma and Sarah, and I thank them profusely.

Academic life can often be lonely, with hours spent reading and writing in isolation. Luckily for me, I have an incredible group of friends that keep me inspired, connected, and sane. This includes Manuel Puppis, Hilde Van den Bulck, Laura Schwartz-Henderson, John Remensperger, Pawel Popiel, Sanjay Jolly, Nora Draper, Andre Cavalcante, Stephen Ninneman, David Nemer, Aynne Kokas, Meredith Clark, Nicholas Mathews, Diane Mathews, Jon Kropko, Cypress Kropko, Augusta Reel, Cathleen ("Clerky") Clerkin, Grey Putnam, Erin Pappas, Mike Laugherty, Jenna Daniel, Megan Daniel, Megan Adams, and Colin Melton. I am grateful for their friendship, their love, and the opportunities we get to share ideas, wine, and laughter.

Family means everything to me, and I am grateful both for the family I was born into and the family I married into. Kim and Bob Stankey and their sons Alex and Nick welcomed me into their home and their lives from the moment I met them. My brother-in-law and sister-in-law, Matt and Nicole Reeves, did the same, as did Susan Weir. I am lost for words when trying to describe how lucky I am to be part of their families. To say "thank you" to my parents Ray and Elaine, my brother Jonathan, and sister-in-law Kelsey (and my nibling, whom, at the writing of these acknowledgments has yet to be born and whom I love with an intensity I did not think I had in me) cannot do justice to my love for them.

I could not have imagined when I began planning this project back in 2016 that in the midst of researching and writing *Farm Fresh Broadband* I would meet, fall in love with, and marry the man of my dreams. Benjamin Reeves simultaneously changed my life and gave me life. For two years he shared the burden of book writing and tolerated my many lapses of attention, hastily scribbled notes, sleepless nights, panicked moments, mood swings, and days without shaving or proper hygiene. I look to him for words of comfort and support, critique and advice, love and encouragement. He has provided these with generous abundance, along with innumerable cups of coffee, glasses of wine, "I love you"s, and "You got this"s. Without him, this book would not have been possible. Without him, I would not be who I am today. He is the wind beneath my wings, the dilithium to my warp core.

To him, and to our misfit family of two Great Danes, Stella and Titan, and the world's best copilot, Tuna the hound, I dedicate this book. All errors in the following text are mine and mine alone.

Abbreviations

A-CAM	Alternate Connect America Cost Model
AFBF	American Farm Bureau Federation
ARRA	American Recovery and Reinvestment Act of 2009
BIP	Broadband Initiatives Program
BOC	Broadband Opportunity Council
CAF	Connect America Fund
CPEC	Critical Political Economy of Communication
CREA	Committee on the Relation of Electricity to Agriculture
CRS	Congressional Research Service
DIY	do-it-yourself
DMCA	Digital Millennium Copyright Act
DRM	digital rights management
DSL	digital subscriber line
DSLAM	digital subscriber line access multiplexer
EFF	Electronic Frontier Foundation
EU	European Union
EULA	end-user license agreement
FCC	Federal Communication Commission
FTTH	fiber-to-the-home
FY	financial year
GAO	Government Accountability Office
GB	gigabyte
Gbps	gigabits per second
GIS	geographic information system
GPS	Global Positioning System
GTFB	Governor's Task Force on Broadband

ICT	information and communications technologies
IoT	internet of things
ISLR	Institute for Local Self Reliance
ISP	internet service provider
KB	kilobyte
LEO	low-Earth orbital satellite
M2M	machine-to-machine
MB	megabyte
Mbps	megabits per second
NBN	National Broadband Network
NBP	National Broadband Plan
NCC	Next Century Cities
NELA	National Electric Light Association
NFU	National Farmers Union
NRECA	National Rural Electric Cooperative Association
NRTC	National Rural Telecommunications Cooperative
NTCA	The Rural Broadband Association
NTIA	National Telecommunications and Information Administration
OECD	Organisation for Economic Co-operation and Development
PRTC	Peoples Rural Telephone Cooperative
R2R	right-to-repair
RDOF	Rural Digital Opportunity Fund
REA	Rural Electrification Administration
RECC	Rural Electric Cooperative Consortium
RoR	rate of return
RRBS	Remote Rural Broadband Systems
RUS	Rural Utilities Service
SHLB	Schools, Health & Libraries Broadband Coalition
TB	terabyte
TVA	Tennessee Valley Authority
UAV	unmanned aerial vehicle
USAC	Universal Service Administrative Company
USDA	United States Department of Agriculture
USF	Universal Service Fund
WIA	Wireless Infrastructure Association
WISP	wireless internet service provider
WISPA	Wireless Internet Service Provider Association

Introduction   Promise, Politics, and Policy: Broadband in Rural America

In 2015, the online tech publication *Ars Technica* reported the story of Nelson Schneider, who wanted to bring high-speed broadband to his farm in Nebraska. As in all areas of life, broadband—or high-speed internet connectivity—has shifted from being a luxury to a necessity in American rural communities, including farmlands. The cost of connection, our farmer was told by his internet service provider (ISP) Windstream, would be $383,500 because the company would need to lay new fiber-optic cables to the farm (Brodkin 2015). This was a financial burden the Schneider farm could not assume. Eventually, the Northeast Nebraska Telephone Company—a small telephone cooperative out of Jackson, Nebraska—offered to bring fiber to the farm for $41,000, an amount more in line with current market prices, where it costs $27,000 per mile to lay fiber-optic cables. Windstream, a publicly traded company and the country's fifth-largest digital subscriber line (DSL) internet provider, offered no explanation for why its estimate was so high. While even the reduced sum of $41,000 may seem outrageous to connect a property to the internet, broadband is as necessary for rural residents as it is to any urban household or university dorm. The problem is getting it there.

The lack of broadband is ubiquitous across rural America and forms the basis of the rural-urban digital divide. In 2020, the Federal Communications Commission (FCC)—the agency charged with setting broadband policy and evaluating broadband deployment—estimated that 22.3 percent of rural America (or 11.37 million people) does not have access to a broadband connection, currently defined as 25 megabits per second (Mbps) download and 3 Mbps upload (FCC 2020a). This is a conservative estimate to say the least. Other studies suggest these figures could be off by upward of 50 percent (Busby and Tanbark 2020; Meinrath 2019; Stegeman 2019). Most alarming, a 2018 study by Microsoft reported a full 50 percent of *all* Americans—162.8 million people—lack access to the internet at broadband speeds (Kahan 2019).

While the exact number of Americans, rural or otherwise, who are unconnected or underconnected continues to elude policymakers, the fact is that in contrast with urban areas, which tend to be well connected and where the digital divide exists because of income gaps and cost (Chao and Park 2020), rural communities tend to lack the infrastructure necessary for connectivity. Making matters worse, rural Americans pay more for broadband and have fewer provider options. Only 19 percent of rural Americans have a choice in broadband provider (Whitacre 2016). The lack of competition drives up prices to amounts unfathomable in urban centers. The lowest-density population areas pay upward of 37 percent more on average for broadband than the densest centers, according to a 2019 report by BroadbandNow (2019a). In Messena, Iowa (pop. 345), for instance, residents pay $145 per month for 25 Mbps using a fixed wireless connection, while the national median is $80 per month (Tibken 2018). In sum, when it's even available, rural Americans pay more for worse connections and fewer options than their urban counterparts.

The challenge of broadband deployment in rural areas is not unique to the United States (see International Telecommunications Union [ITU] 2019; Organisation for Economic Co-operation and Development [OECD] 2018; Taylor and Middleton 2020; J. van Dijk 2020). According to the ITU (2019), 53.6 percent of the world's population has access to a fixed broadband connection. While a notable achievement, only 14 percent of rural areas in the Global South have an internet connection (compared with 46 percent of urban areas in the Global South) (Alliance for Affordable Internet 2020). As Judith Mariscal (2020) recently observed, "The original objective . . . being a mechanism to deploy broadband to remote and underserved areas is a target that has been elusive for many countries" (134). Countries of a comparable geographic size to the United States, such as Canada and Australia, have grappled with the challenge of connecting their vast "remote rural" places (B. Howell and Potgieter 2020; Taylor 2018). Canada particularly has struggled to connect its northernmost communities (Ruiz 2014). Australia has attempted to build the publicly funded National Broadband Network (NBN) to connect the country, although it has been mired in delays and controversy (B. Howell and Potgieter 2020).[1] Other geographically smaller but more populated countries, namely the United Kingdom and Germany, have also struggled to reach the "final few" (Philip et al. 2017) and broadband "notspots" (Townsend et al. 2013). Over 164 countries have national broadband plans to solve the digital divide (ITU 2019), as do the ITU (2019) and OECD (2018). In June 2020, the secretary-general of the United Nations called the digital divide "a matter of life and death" amid the COVID-19 pandemic (Guterres 2020). There is a universal question here: How do we connect the rural and remote areas of the world with broadband? This question is that much more pressing during a pandemic, when those with the means to do so have migrated online and those without are forced further into a state of what Roberto Gallardo and Cheyanne Geideman (2019) call "digital distress."

Back in the United States, the existence of the rural-urban digital divide has not been lost on the federal government. In 2017, newly elected president Donald Trump stood before the annual meeting of the American Farm Bureau Federation (AFBF) in North Carolina and pledged his support to fixing the problem of rural broadband:

> It shall therefore be the policy of the executive branch to use all viable tools to accelerate the deployment and adoption of affordable, reliable, modern high-speed broadband connectivity in rural America, including rural homes, farms, small businesses, manufacturing and production sites, tribal communities, transportation systems, and healthcare and education facilities. (Trump 2018)

Following his remarks, the president signed two highly touted, but ultimately benign, executive orders: one easing regulations for telecommunications companies to make use of government land for cell phone towers and poles and the other promising to reduce paperwork for companies wanting to serve rural areas with broadband (Trump 2018). While these orders may eventually help broadband providers to deploy their services more efficiently, neither incentivizes providers to come to rural America in the first place. That takes money: lots of it.

The federal government subsidizes broadband to the amount of $9–$10 billion a year.[2] About $6 billion of this sum is devoted specifically to *rural broadband* deployment, which is seen as the most expensive form of connection given the sparse, and vastly spread out, population. The money comes from a mixture of loans, grants, and cross-industry subsidies and hails from four programs operated by the FCC under the auspices of Universal Service Administrative Company (USAC); five programs from the Rural Utilities Service (RUS) (a division of the US Department of Agriculture [USDA]); and a handful of other federal agencies such as the National Telecommunications and Information Administration (NTIA, a division of the Commerce Department), the Department of Health and Human Services, the Department of Education, the Department of Defense, along with numerous state governments. These programs all operate under the same policy goal—to achieve universal service—and they exist to incentivize private companies to deploy broadband where there is no market rationale to do so.

To be sure, $6 billion is nothing to scoff at. While it does not reach the levels necessary to complete the promise of universal high-speed broadband (approximately $61–$80 billion for fiber-to-the-home (FTTH) everywhere [Crawford 2019; de Sa 2017; Engebretson 2018a]), when added up year by year, it should get us close. The truth, however, is that we are nowhere close to bridging the digital divide facing rural Americans (and low-income, minority, or Indigenous Americans, for that matter). Instead, the divide is growing as telecommunications companies now speak of deploying the latest 5G and gigabit service to major cities while many rural Americans still live "in the dial-up age" (Levitz and Bauerlein 2017).

Farm Fresh Broadband

The stories of the Schneider farm in Nebraska and Trump in North Carolina capture this book's hallmark themes. First, they acknowledge the importance of broadband for rural communities and the fact that rural Americans desire this connectivity. Second, they speak to the failure of the private market, especially large telecommunications companies, to deliver this vital utility equally and efficiently to all Americans. Third, they intimate how policy mechanisms designed to encourage rural broadband deployment have failed, despite the hype. Fourth, in the case of the Northeast Nebraska Telephone Cooperative, which eventually wired the Schneider farm, they highlight the importance of local companies and cooperatives to rural broadband and rural communities. Fifth, they speak to the timeliness of this topic. Connectivity, or lack thereof, is the most pressing communication issue and perhaps the most pressing infrastructure issue facing rural Americans today. A full 24 percent of rural Americans say that broadband access is a major problem in their community (Anderson 2018), and in the summer of 2019 no fewer than five 2020 Democratic presidential candidates released plans for rural broadband. Senator Elizabeth Warren (2019), for instance, writes, "One of the best tools for unlocking economic opportunity and advances in health care, like telemedicine, is access to reliable, high-speed Internet. In the twenty-first century, every home should have access to this technology—but we're not even close to that today." Warren and the other presidential hopefuls joined a multiyear conversation in the mainstream media that has seen dozens of op-eds, press articles, and studies exalting the importance of broadband to rural America and extolling the need to end the rural-urban digital divide.[3] This conversation was kicked into high gear in the winter of 2020, when the world was in the grips of the COVID-19 pandemic. In the United States, the pandemic cast in sharp relief the pervasiveness and destructiveness of the digital divide once the country was forced to live, work, shop, seek health care, and communicate with loved ones online (Holpuch 2020). Concerns abounded for those who did not have access to broadband, especially young students (Fishbane and Tomer 2020) and the unemployed (Sohn 2020), with some doubting whether American broadband networks could support this migration online (Alba and Kang 2020).

Farm Fresh Broadband recalls these conversations and advances them by analyzing how US policymakers are tackling the rural-urban digital divide. This is a book about failures and successes: failures of policies and markets to deliver broadband and successes of communities to come together and solve their broadband deficits. In the following pages I analyze the complicated terrain of rural broadband policy, with examples drawn from the rural and agricultural communities of the American Midwest. Despite academic and popular press attention to various issues related to rural broadband— economic development, health care, education, mapping—the United States lacks a systematic and comprehensive assessment of rural broadband policy, technology, markets, and stakeholders.[4] Rectifying this situation, I explore different business models, emphasizing the importance of cooperatives to provide rural broadband. I contemplate the technologies of rural broadband—from the "futureproof" fiber to tried, tested, and tired (and slow) DSL and the rollout of the advanced cellular networks of 4G, LTE, and 5G. I examine the uses of rural broadband, focusing on precision agriculture (application of information and communications technologies [ICT] to agriculture). Most importantly, I analyze the policies of rural broadband, interrogating issues of regulatory jurisdiction, authority, legitimacy, subsidy, and mapping. The key question guiding the book is: Where is the $6 billion a year the federal government distributes for rural broadband going?

Before I can begin to answer that question, I need to establish some common vocabulary: "broadband," "rural broadband," and "rural." I will then introduce my research questions, the theories and methods that grounded and framed this research, and the argument that defines the book. Throughout the rest of this introductory chapter and the entire book, I encourage urban readers to think about their own experiences using broadband: How frustrated do you get when you can't upload your taxes, can't stream your favorite show or Skype with your parents, can't download paperwork, make a call, or use Google Maps? These are vexing but rather short-lived frustrations for urban middle-class Americans—a reset of the wireless router, finding a Wi-Fi hotspot, or getting closer to town will usually correct these inconveniences. For over 11 million rural Americans, however, these frustrations define their digital lives—and this is so much more than just Netflix. As one respondent told me,

> So, it's not just you didn't get your HBO last night and see *Game of Thrones*; you didn't plant your rows in a more precise way that would allow you much greater returns at the end of season to pay back the loans on the hundreds of thousands of dollars towards equipment you just purchased. (A. Brozana, personal communication, 1/18/18)

Contrast this with what someone in a small southwest Minnesota town told me about getting high-speed internet:

> It changes everything. If you walk outside, it changes everything. But you look around, it changes nothing. You don't see it. Everything's the same. But it changes everything inside these businesses and homes. (M. Erickson, personal communication, 7/25/18)

This is the change *Farm Fresh Broadband* seeks. For urban readers, I want to convince you of the importance of this issue. For rural readers, I hope you see yourself in these forthcoming pages.

Broadband and the Rural-Urban Digital Divide

"Broadband" is a generic term used to describe high-speed internet connectivity.[5] It is defined by speed, rather than technology. Currently, the FCC—the agency with regulatory authority over broadband—defines it as an "always on" connection with a download speed of at least 25 Mbps and an upload speed of 3 Mbps (commonly depicted as "25/3"). This threshold was established in 2015, when the FCC increased it from 4/1. At 25/3, you can stream a Netflix show on your Apple TV while browsing Instagram on your smartphone and encounter little buffering or delay. It is not, however, enough for multiple household members to stream and Instagram and game at the same time. And speed is not just speed, it's also delay (known as "latency") and cost, since high speeds may be unattainable when they come with a high price tag (Chao and Park 2020). As broadband expert Jonathan Chambers (2018) explains,

> A speed of 10 Mbps/1 Mbps with only a 10 GB [gigabyte] month data cap is not broadband. The average household use of broadband has grown a hundredfold over the past decade to 200 GB per month. Similarly, 10 Mbps/1 Mbps with a latency of 750 milliseconds is not broadband because voice and other key applications require low-latency networks. Nor is 10 Mbps/1 Mbps broadband if it is priced at $100 per month.

Broadly, therefore, broadband is

> provided by a series of technologies (e.g., cable, telephone, wire, fiber, satellite, mobile, fixed wireless) that gives users the ability to send and receive data at volumes and speeds necessary to support a number of applications including voice communications, entertainment, telemedicine, distance education, telework, ecommerce, civic engagement, public safety, and energy conservation. (Congressional Research Service [CRS] 2019a, i)

Definitions of broadband range across the world (Souter and van der Spuy 2019). For those countries that define broadband by speed (rather than qualitatively by use) the FCC's definition of broadband seems unambitious. Canada has set baseline broadband speeds at 50/10 (Canadian Radio-Television and Telecommunications Commission 2020). Germany has a minimum broadband speed of 50 Mbps for previously underserved communities (European Commission 2020). In the United Kingdom and European Union (EU), broadband (or "fast broadband") is defined as 30 Mbps download (House of Commons Library 2019). The EU has a goal of 100 percent broadband penetration of 30 Mbps for member countries by the end of 2020 (OECD 2018). For comparison, the FCC's 2010 National Broadband Plan (NBP) had a goal of 100 Mbps download to 100 million homes (FCC 2010). In fairness to the United States, the ITU currently defines broadband "as everything greater than or equal to 256kbps" which, according to a UNESCO report "would not be considered broadband in most communications markets now" (ITU 2019; Souter and van der Spuy 2019, 198). For clarity and consistency, and given the US focus of this book, I use (and critique) the FCC's

25/3 definition of broadband, but it is nevertheless important and useful to see how the United States compares in its broadband definitions and goals.

For the better part of three decades, scholars, policymakers, and industry watchers have acknowledged that those who lack internet access form part of what is known as the digital divide (see Hoffman, Novak, and Schlosser 2000; J. van Dijk 2020). Communications scholar Jan van Dijk (2020) suggests that the term *digital divide* was first coined by *Los Angeles Times* journalists Jonathan Webber and Amy Harmon in a July 1995 article.[6] Van Dijk (2020) goes on to explain that there are three levels of the digital divide. The first level is about connectivity—the haves and have-nots; the second is about skills and usage; and the third is about the outcomes of the digital divide. This book focuses on the first level, taking note of the fact that we have yet to solve the issue of access here in the United States.

Since the 1930s, lawmakers have understood that rural America needs to be on par with its urban counterpart when it comes to the trappings of modern life. At that time, it meant electricity, automobiles, and telephony (Kline 2000). As Franklin Roosevelt said in his 1932 presidential campaign: "[Electricity] is no longer a luxury. It is a definite necessity. . . . It can relieve the drudgery of the housewife and lift the great burden off the shoulders of the hardworking farmer" (qtd. in Brown 1980, 34). Today, we say the same about broadband:

> Bringing ubiquitous and affordable broadband service to rural America will improve the quality of education, healthcare and public safety in rural America, among other benefits. On a larger scale, ensuring that all Americans, including those in rural areas, have access to such services will help improve America's economy, its ability to compete internationally, and its unity as a nation. (Copps 2009, 8)

Since its identification in the mid-1990s, the notion of the digital divide has evolved from one simply denoting access, to incorporating "digital inclusion," or access, plus everything else (NTIA 2000; Rhinesmith 2016; Roberts et al. 2017; J. van Dijk 2020). As the National Digital Inclusion Alliance (2017) explains,

> Digital Inclusion refers to the activities necessary to ensure that all individuals and communities, including the most disadvantaged, have access to and use of Information and Communication Technologies (ICTs). This includes 5 elements: 1) affordable, robust broadband internet service; 2) internet-enabled devices that meet the needs of the user; 3) access to digital literacy training; 4) quality technical support; and 5) applications and online content designed to enable and

encourage self-sufficiency, participation and collaboration. Digital Inclusion must evolve as technology advances. Digital Inclusion requires intentional strategies and investments to reduce and eliminate historical, institutional and structural barriers to access and use technology.

This definition captures not only access and availability of connections but also affordability, network stability, access to devices, and digital literacy. As Jonathan Sallet (2020) reminds us, broadband is not about policies or technologies; it is about people.

The digital divide is also not an exclusively rural-urban problematic, but one defined and explained by inequality. Certainly, there is the rural-urban digital divide, of which this book speaks, but there are also digital divides present among low-income Americans (Anderson and Kumar 2019; Chao and Park 2020), African Americans (Hoffman and Novak 1998; Moran and Bui 2019; Turner 2016), tribal nations (Duarte 2017), and newcomers/immigrants (Fairlie et al. 2006). These divides also intersect. A study by Free Press (Turner 2016) revealed how, when broken down by race, 19.7 percent of white rural Americans lack access to broadband (defined at the agonizingly slow speed of 3 Mbps in this study), while 32.3 percent of rural Hispanics, 21.8 percent of rural African Americans, and 43.2 percent of rural Native Americans "are completely unserved by any wired ISP even at that relatively low speed" (10). When the speed threshold is increased to 25 Mbps download, this intersectional digital divide of race and community grows, with 52.1 percent of the rural Hispanic population, 44.6 percent of the rural African American population, and 67.1 percent of the rural Native American population unconnected (Turner 2016).

The reasons for these digital divides are many, with income, cost, age, interest, location, and race prevalent among them (Anderson et al. 2019). The fact of the matter is, however, that digital divides exist because of a failure of the private market to provide this essential service and a failure of policy to create the necessary market conditions or, better yet, categorize broadband as a basic necessity or human right. As Marissa Duarte (2017) explains in her brilliant study of broadband deployment among Indigenous peoples:

In the United States, the national telecommunications infrastructure is one of the only major infrastructures—the others are the interstate transportation system and electric power grid—that has been primarily market driven. The net effect is that urban and semi-urban locations, and locations neighboring major highways, receive the most robust and competitively priced ICT services—landline, wireless, fiber-optic cable, and satellite connectivity—while many tribes lack the infrastructure for basic phone service. (91)

While the United Nations, the EU, and a host of countries have moved beyond thinking about broadband as a pure economic good and frame it as a human right (Jasmontaite and de Hert 2019) or "basic service" (Canadian Radio-Television and Telecommunications Commissioner 2016), the United States is stuck in a "market ontology" (Pickard 2015): broadband is a luxury for those who can afford it and where the market sees the highest return on investments. This, apparently, does not include rural America.

The lack of access to advanced telecommunications in rural America contributes to what is known as the "rural penalty." The rural penalty describes the material and figurative costs paid by rural residents and businesses for their geographic distance from the centers of commerce and culture (Hite 1997; Nicholas 2003; E. Parker et al. 1989). The costs include "small markets, high transport costs, and physical isolation" brought on by the "tyranny of distance" (Kandilov and Renkow 2010; MacDougall 2014). Advanced telecommunications, first in terms of the telephone and now in terms of broadband, are thought of as the "distance-killing technolog[ies]" necessary to correct for the rural penalty (Hindman 2000; Malecki and Moriset 2003; E. Parker et al. 1989). Unfortunately, the time-space dialectic still holds on to life in rural America despite efforts at its erasure by both capitalism and telecommunications. That these rural-urban divides persist into the broadband age suggests, therefore, that the digital divide is not a natural or technological problem but rather a problem of political economy. As Edwin Parker et al. observed in 1989, "There is no inherent technical reason for the historic 'rural penalty' of geographical remoteness. Yet the rural penalty persists because policies affecting telecommunications and economic development have not kept pace with the times and taken sufficient account of changing rural economic needs" (5). I furthermore suggest the digital divide persists because it is in the interests of the largest telecommunications companies to preserve it. Akin to the pharmaceutical industry, where it is better for pharmaceutical companies' bottom line to treat rather than cure a patient, it is better for telecommunications companies' bottom line to keep the digital divide intact and offer incremental improvements rather than make the changes necessary to end the digital gaps.

According to the FCC's *2020 Broadband Deployment Report*, 5.6 percent of Americans, including 1.5 percent of urban Americans, 22.3 percent of rural Americans, and 27.7 percent of residents on tribal lands, do not have access to broadband speeds at 25/3 (FCC 2020a; see table 0.1). In terms of adoption, 63 percent of rural Americans have a home subscription, while 15 percent of rural adults say they never go online (Perrin 2019). When availability is not a problem, cost is a major determinant for those who do not subscribe (Chao and Park 2020; Horrigan and Duggan 2015). This holds true for both urban and rural areas (Horrigan 2020; Levin and Downes 2019). These unconnected and underconnected communities are in "digital distress," defined as those households without internet access at all, those with only a cellular data subscription, those without a fixed broadband connection, those relying only on mobile devices, and those without a computer (Gallardo and Geideman 2019). These communities "have a harder time using and leveraging the internet to improve their quality of life due to the type of internet subscription or devices owned" (Gallardo and Geideman 2019).

The FCC numbers and percentages depicted in table 0.1 tell a tragically incomplete story as the Commission relies on a flawed data collection methodology. As a result, it has dramatically overestimated how many Americans have access to broadband (see chapter 2). These numbers also assume parity among the technologies of access when this is anything but the case. Not all internet connections are created equal and not all deliver what Sallet (2019a) calls "high-performance broadband." Just like with price and availability, rural Americans lack access to the advanced networks of tomorrow.

Technologies of Access

There are six ways Americans access the internet at broadband speed, which can be divided between wired and wireless connections.

Wired Connections

DSL uses the enhanced copper lines of the legacy telephone companies like AT&T, Frontier, CenturyLink, and Windstream.[7] They are some of the oldest technologies providing broadband to America and a mainstay throughout rural communities. In fact, DSL has the largest rural footprint of any broadband technology in the United States, available to 75.7 percent of rural housing units (Gallardo and Whitacre 2019) (see table 0.2).[8] While near-omnipresent, DSL cannot transport data at the speed of other wired

Table 0.1

Percentage of Americans with access to fixed terrestrial broadband (population in millions)

	2014		2015		2016		2017		2018	
	Pop.	%	Pop.	%	Pop.	%	Pop.	%	Pop.	%
United States	284.246	89.4%	287.853	89.9%	263.373	91.9%	304.405	93.5%	308.913	94.4%
Rural areas	37.174	60.3%	38.271	61.5%	42.677	67.8%	46.960	73.6%	50.99	77.7%
Urban areas	247.072	96.4%	249.582	96.7%	253.695	97.7%	257.446	98.3%	258.814	98.5%
Tribal lands	2.245	57.1%	2.290	57.8%	2.520	63.1%	2.727	67.9%	2.921	72.3%
Pop. evaluated	317.954	100%	320.289	100%	322.518	100%	325.716	100%	327.167	100%

Source: FCC (2019a, 2020a).

technologies and has median download and upload speeds of 10/1 (Gallardo and Whitacre 2019). These, of course, do not meet the FCC's definition of broadband. As speeds are asymmetric, download speed far outpaces upload speed, to the detriment of businesses, Skypers, and gamers alike. Moreover, DSL suffers from high signal attenuation after about three miles, meaning that the farther you live from a network exchange (a digital subscriber line access multiplexer [DSLAM]), the slower your internet connection will be.

This brings us to another problem: many DSL providers refuse to upgrade or repair their lines in rural areas beyond regulatory commitments (*Fauquier Times* 2019). In 2018, after its merger with backbone provider Level3, CenturyLink, one of the largest rural broadband providers and the seventh-largest telecommunications company in the country, announced it would focus primarily on high-density (i.e., metro) areas, with the exception of where it had secured subsidy. Similarly, AT&T is trying to move its rural customers to wireless, while Verizon has "been ignoring the rural parts of their network for literally decades" (Dawson 2018b). Indeed, Verizon offloaded most of its rural DSL network to Frontier some years ago, and Frontier has been following this trajectory of infrastructure neglect (Gonzalez 2018). The refusal of major telecommunications companies—AT&T, Verizon, CenturyLink, Frontier—to maintain and upgrade their DSL lines is but one example of the market and policy failures within rural broadband. FCC policy, however, continues to privilege, and in many cases protect, incumbent telecommunications providers through guaranteed subsidy and favorable regulatory conditions while the providers themselves refuse to upgrade their networks and use the speeds mandated by their subsidy (10/1) as a ceiling rather than a floor.

While DSL is dominant in rural America (see table 0.2), coaxial cable dominates in urban communities. Cable is the most ubiquitous broadband technology in America, accessible to 88.2 percent of households (including 55.1 percent of rural households) (Gallardo and Whitacre 2019). Comcast is the largest cable broadband provider in the country (64.8 million subscribers), followed by Charter Spectrum (57.2 million subscribers) and a host of other local cable companies. Connected via coaxial cable, cable broadband provides asymmetrical service, with thoroughly decent download speeds but slow upload speeds. Gallardo and Whitacre (2019) report that in the US the median cable download speed is 400 Mbps and the median upload speed is 20 Mbps. This asymmetry privileges consumption over production

Table 0.2
The footprint of fixed broadband technologies in rural America

	DSL	Fixed wireless	Cable	Fiber-optic
Percentage of rural housing units passed	75.7	43.2	55.1	16.5
Median download speed (Mbps)	10	12	300	1,000 (or 1 Gbps)
Median upload speed (Mbps)	1	3	20	150

Source: Gallardo and Whitacre (2019).

and consumers over businesses. As was explained to me, "download is consumption, upload is production" (M. Erickson, personal communication, 7/25/18). Susan Crawford (2019) goes further in explaining the difference:

> People who say, "who needs symmetric access?" today are just like those who said "who needs electricity for anything but lightbulbs?" a hundred years ago. Because symmetric high-capacity networks don't yet exist in the giant U.S. market, the applications to use them, our twenty-first-century analogs to electric kitchen appliances, also don't yet exist. (50)

In terms of deployment, the major cable companies do not trek into rural America. As of 2018, for instance, Comcast had no presence in Iowa, Nebraska, North Dakota, South Dakota, Wyoming, or Oklahoma, while Charter Spectrum had no presence in North Dakota, South Dakota, Iowa, Oklahoma, Arkansas, and Utah (C. Mitchell and Trostle 2018). That said, many rural communities are served by small cable companies. Luverne, Minnesota (pop. 4,619)—featured prominently in chapter 4—is served by cable providers Vast Broadband and Mediacom, which advertise speeds of 200 Mbps (BroadbandNow 2019b).

Fiber-optic broadband is the gold standard of internet connectivity, garnering the moniker "future proof" because the glass wires that comprise a fiber-optic cable can handle near-unlimited amounts of data. As Crawford (2019) notes, "A single fiber-optic cable can carry the entire weight of data on the internet" (22). With data transmitted by light waves over glass filaments, FTTH can give residents over 1 Gbps symmetrical service—speeds vitally important for modern businesses, which tend to upload much more than download. Farmers equipped with precision agriculture equipment, for instance, may upload terabytes of information (high-resolution photos, aerial shots from drones, soil measurements, etc.) a day, and anything short of fiber would create enormous delays. As one agricultural drone manufacturer in Winnebago, Minnesota, told me, his business would not have been possible without fiber broadband. With fiber, only the electronic encoding equipment and not the wires themselves will need upgrading, meaning that only fiber will be able to adjust for the coming internet of things (IoT)—a world of connected cars, combines, toasters, televisions, tractors, Fitbits, and even holographic communication (Baumgartner 2018; Crawford 2019).

The installation of fiber, however, is very costly. Like DSL or cable, fiber-optic cables can either be buried or strung along poles. Both options, however, are prohibitively expensive. The average cost of deploying fiber-optic cable is $27,000 per mile, but this varies significantly between rural and urban areas and between areas with and without existing infrastructure (Aman 2017). The Schools, Health and Libraries Broadband Coalition estimates that new aerial construction would cost $51,000 per mile, while new underground trenching would cost $86,000 (CTC Technology & Energy 2018). This makes it all the more difficult to bury or string fiber in rural communities with much less density per square mile. Fewer people and more space make this impossible without grants and subsidies. That said, when ISPs or communities themselves invest in fiber, the benefits are obvious and immediate. Such was the case in McKee, Kentucky, county seat of Jackson. Jackson is one of the most economically depressed counties in the state, but one with FTTH thanks to their co-op telephone company, the wonderfully named Peoples Rural Telephone Cooperative (PRTC). As its president, Keith Gabbard, said to me,

> Of course, this is a huge investment. In the long run, [it would] be less maintenance, wouldn't have to be rebuilding all the time, and we thought we could provide better products. As we got farther into it, we realized, okay, this could be a game changer in education, in . . . development, in healthcare. It could improve our economy and . . . we [have] definitely seen some changes, and I think even we didn't realize how big of a deal it would be back when we started ten years ago. (personal communication, 1/18/18)

Between wired and wireless connection is "fixed wireless"—a network where a DSL or fiber-optic line is passed to a tower and an antenna sitting on that tower beams a wireless signal to the home (or to another tower).[9] Provided by companies known as WISPs (wireless internet service providers), fixed wireless systems offer speeds comparable to those generated by DSL and have been growing in popularity in rural communities because their signals can cover some distance (8–10 miles). The networks are cheaper than laying fiber to the home and may come with a fiber "ring"—usually a series of fiber-optic cables connecting towers and anchor institutions like schools, libraries, and government buildings. Many providers are pursuing this as a stopgap before securing funding for FTTH (Carlson and Mitchell 2016). A recent article on BroadbandNow praises WISPs for "actually working to bridge the [digital] divide . . . bringing reasonable speeds to areas that have been left behind by technological progress" (Cooper 2018b). Fixed wireless relies on either unlicensed or licensed spectrum to deliver its transmission. During the COVID-19 pandemic, the FCC gave thirty-three WISPs access to the 5.9 GHz band, normally reserved for short-range communication, for 60 days to "meet a surge in demand for residential fixed broadband access" (Alleven 2020a). While this is only a temporary solution, FCC chairman Ajit Pai has pushed for a portion of the 5.9 GHz band to be made available as unlicensed spectrum for fixed wireless use (Hill 2020). At the same time (spring 2020), the FCC released spectrum in the 6 GHz band for unlicensed use. While the primary purpose of this is for enhanced Wi-Fi, it will also significantly benefit rural WISPs (Engebretson 2020a). The president of the Wireless Internet Service Providers Association (WISPA), Claude Aiken, praised the FCC's April 2020 decision:

> We hope that for small rural innovators such as WISPs, the power levels in the order mean they'll have more bandwidth to grow their business, develop new transmission techniques such as fixed 5G and, perhaps most importantly, bridge the rural digital divide, providing millions of rural Americans new options and tools to obtain the same prosperity and safety their urban counterparts expect and enjoy. (qtd. in Engebretson 2020a)

In the summer of 2020, the FCC also commenced an auction for licenses on the 2.5 GHz band for rural tribal broadband (Early 2020). This is an important nod toward tribal cyber sovereignty (Duarte 2018) and a decision that increases the profile of fixed wireless even more.

In addition to traditional fixed wireless, experiments have begun that use television white spaces (the "gaps" between television frequencies) for wireless broadband. In 2017, Microsoft announced that it would begin such fixed wireless experiments (Smith 2018). Through its Airband Initiative, Microsoft sought partnership with twelve rural communities and private ISPs to test the hypothesis. Their goal is to connect 3 million rural Americans by 2022.[10] Canada has also experimented with broadband via television white spaces through its Remote Rural Broadband Systems (RRBS) program. As Gregory Taylor of the University of Calgary explained in a 2019 article, the RRBS held tremendous promise for connecting Canada's rural communities:

> It encourages and supports new entrants into the wireless broadband sector; it makes use of spectrum that is by and large idle; it explicitly seeks to expand service into underserved areas and the signal provided by these frequencies offers strong propagation qualities, with the ability to penetrate a common obstacle in rural Canada: trees. (Taylor 2019, 744)

In addition to trees, RRBS had promise for farmland as it could cover up to twenty kilometers. While the RRBS was canceled before it got off the ground, experiments with television white spaces are gaining in popularity the world over (Taylor and Middleton 2020).

There are drawbacks with fixed wireless, however. Many towers, for instance, are not equipped with a fiber connection, meaning the speeds they can provide their users are minimal because they are relying on a DSL or microwave backhaul. Fixed wireless requires line of sight (the receiving antenna must be in view of the tower), so the farther you are from the tower, the weaker your connection will be. There are also frequent equipment replacements, the cost of laying the fiber to the tower, and, of course, issues caused by inclement weather (CCG Consulting 2014). If we take the example of Luverne, Minnesota, there are two fixed wireless providers serving the area. MVTV Wireless's fastest speed is 5 Mbps and Lismore Wireless comes in at 20 Mbps (BroadbandNow 2019b). Neither, therefore, meet the FCC's definition of broadband and both are expensive ($74.95 and $67.95 per month, respectively).

Wireless Connections

In a 2016 debate over the definition of broadband speed, the FCC contemplated whether cellular internet was the equivalent of, and therefore could replace, wired connections. After all, 83 percent of Americans and 71 percent of rural Americans own an internet-enabled smartphone (Perrin 2019). But mobile broadband simply cannot compete with a fixed connection. Not only would it be exceptionally difficult to fill out forms or write school essays on one's mobile device, but cellular customers must also deal with network interruptions, data caps, cost, and subpar download and upload speeds. Despite the maps we see from mobile providers like AT&T and T-Mobile, fourth-generation coverage (known as 4G LTE)—the mobile connection necessary to stream content—is far from universal (Grubesic and

Mack 2017).[11] Indeed, in 2019 it was revealed that the major wireless providers had exaggerated their network coverage by upward of 40 percent (FCC 2019d).

A major problem with the belief that cellular and fixed connections are equivalent is the disregard for data overages. To rely exclusively on cellular networks, as rural residents often do and are being told to do more, means living your entire digital life on your phone or using your phone as a "hotspot." This means butting up against data overages, which can be hugely expensive. In 2017, for instance, Verizon severed contracts with 8,500 customers, all in "rural" states, because of data overages (Brodkin 2017a). The telecom giant claimed that "roaming charges have made certain customer accounts unprofitable for the carrier" (Brodkin 2017a). While the FCC has ruled that fixed and mobile broadband are not fungible, this decision has not stopped carriers from trying to move their rural customers onto or off of mobile-only plans.

Despite the tremendous expense and inconvenience, mobile broadband is a lifeline for many in rural America. According to the Pew Research Center (2019), 20 percent of rural Americans are "smartphone only" internet users. Another popular mobile platform is the standalone hotspot such as Verizon's Jetpack product. These products can be also expensive and come with monthly data caps and hefty overage charges ($10 per GB extra). Telecommunications consultant Doug Dawson (2020) has reported that some "rural households . . . spend $500 . . . per month or more for the hotspot plan." Nevertheless, mobile and hotspot broadband have been essential for those without fixed broadband, even more during the shelter-in-place advisories that accompanied the COVID-19 pandemic (Khazan 2020)

Like cellular, satellite internet has been championed as a panacea for rural America. Looking at the FCC's broadband map (see figure 2.1), one cannot help but notice that almost every community appears to be served by at least two broadband providers. This is because the FCC counts satellite internet as a viable competitor to wireless and wired broadband. They are wrong to do so.

Satellite internet, while covering 99 percent of the country, is notoriously slow, delayed, expensive and comes with miniscule data caps (Dan Koeppel [2019] of *Wirecutter* reports data caps ranging from 10BG to 100GB, in comparison the average daily data usage per user on fixed networks is 15.46 GB [Britt 2020]).[12] The FCC (2018d) reports an average advertised speed of 15–25 Mbps download and 1–3 Mbps upload for satellite internet. While both Viasat and HughesNet, the two largest satellite internet providers in the United States, upgraded their systems in an attempt to meet the FCC's 25/3 threshold, satellite broadband is also plagued by high latency. Brian Whitacre et al. (2018) estimate that if satellite internet is taken off the table as a "reasonable replacement," then "the number of Americans without access to 25/3 speeds would nearly double. . . . 75 percent of this population is classified as rural by the FCC." By claiming satellite as "broadband," however, the FCC can lay claim to having more Americans connected to the digital grid. The implications for rural America are tremendous as wired providers may avoid serving these areas, or making upgrades in these areas, because they are already "served" by a satellite provider.

5G: The Next Generation?

While I was writing the first iteration of this introduction (fall 2018), the debate around the next generation of mobile connectivity—5G—quickly ramped up. 5G represents stronger mobile broadband that delivers the bandwidth necessary to handle the IoT—those everyday household objects, in addition to our laptops, smartphones, and tablets, that will connect to the internet. Think thermostats, Alexas, fridges, and cars. 5G can be 10–100 times as fast as 4G and is powerful enough to, in theory, replace your home fixed broadband network (Thompson and Vande Stadt 2017).

5G operates on three different frequency bands, each with different properties. The top-of-the-line band is known as "millimeter wave" or "high-band." It is located above 24 GHz on the electromagnetic spectrum. This is the 5G that mobile providers promise us and the band that the United States touts. It is the 5G of the IoT, machine-to-machine (M2M) communication, and ultrafast connections, with the potential for over 1 Gbps symmetrical service. Verizon is rolling out its networks on high-band frequencies, as is AT&T for parts of its network (Rizzato and Fogg 2020). The problem is that because the bandwidth needed to fuel millimeter wave 5G is so intense, signals cannot travel far compared with 4G LTE, which can travel upward of tens of miles depending on topography (Holma and Toskala 2011).[13] Indeed, a millimeter wave 5G signal is so weak that it can only travel about 800–1500 feet before requiring a signal boost. Multiple "small cells"—boxes about the size of a pizza box attached to telephone poles and street signs—are needed to keep a driverless car, or marathon runner using a Fitbit, constantly connected. These small cells also require a fiber backbone to run the signals back, necessitating even more digging and laying of fiber. According to one estimate, connecting the country via 5G may cost more than $300 billion (compared to $61–$80 billion to connect the country to fiber) (Engebretson 2018a; Goldman 2018).

Mid-band sits between 3 and 5 GHz. Its signal travels farther than high-band (several miles), but the bandwidth and connection speeds are not as strong (J. Horwitz 2019). Given the limited geographic range of high-band, many countries outside the United States have chosen a policy position of mid-band deployment as a compromise between coverage and performance (O'Donnell 2019). The European Commission (2019) recently announced plans to harmonize the 3.6 GHz band to ensure Europe-wide mid-band 5G (see also Kinney, 2019). According to the European Commission, "the 3.6 GHz band has been identified as the primary pioneer band for 5G in the European Union" (2019). In the United States, Sprint based its 5G network on mid-band frequencies. Nevertheless, mid-band auctions have been hard to come by in this country, making mid-band 5G scarce.[14]

Low-band 5G sits below 1 GHz, usually resting between 600 and 900 MHz. Out of all the 5G bands, it can travel the farthest (hundreds of square miles according to T-Mobile; see Engebretson 2018c), making it ideal for rural deployment. This is the strategy of T-Mobile for its 5G network. The downside is that the user experience will be nothing compared to high-band 5G. Indeed, the user will probably not experience a significant difference in their mobile connectivity between 4G LTE and low-band 5G. Still, for rural Americans, where 4G remains spotty, it will be an improvement (Salter 2020).

Policymakers and telecommunications companies are touting 5G as the solution to rural broadband and the digital divide. In 2019, for instance, T-Mobile and Sprint pledged full rural 5G coverage if they were allowed to merge (Engebretson 2019c). Some have gone so far as to suggest that 5G will make the need for a wired internet connection obsolete (e.g., Rysavy 2018). While this may be true for urban homes and business—something Verizon is already piloting in Michigan, Texas, Florida, California, and Washington, DC—it falls victim to the same fallacies as other conversations on rural broadband: accessibility and geography. High-band 5G is so geographically limited that it will only be viable in densely populated areas—cities and large towns. Rural America is too sparsely populated and too vast for a company to justify small-cell deployment. With a range of 800 feet, imagine how many cells would be required for a 1,000-acre farm! The sparse population of rural America leaves little economic incentive for the champions of 5G—AT&T, Verizon, T-Mobile, and Sprint—to upgrade their rural networks. For high-band 5G, it is estimated that every customer in a rural community would require their own small cell, at a cost of $30,000 to $50,000 per customer (Thompson and Vande Stadt 2017). Low-band 5G, which is what is currently being deployed in rural America, does not bring with it promises of ultrafast connectivity, thus exacerbating the digital divide. While urban America gets the network of the future, rural America is stuck with a network that is simply "good enough."

Looking through the prism of the technologies of access that comprise broadband deployment, it can be seen that rural Americans have been left behind, not only by markets and policies, but by technological innovations as well. Many residents are forced to use antiquated technologies like DSL or 3G, purchase prohibitively expensive satellite or mobile plans, drive to McDonald's for a Wi-Fi connection, or, in many instances, simply forgo their digital dreams while waiting in a Beckettian nightmare for the day when fiber or 5G comes to town.

The Five Pillars of Rural Broadband

Rural broadband, and indeed the entire history of rural communications, is defined by what economists call "market failure." Market failures exist when the private market is unable or unwilling to provide for a socially important good because of a lack of return on investment (Ali 2016; Bator 1959; Pickard 2013). Rural broadband, by definition, is a market failure, and the lack of broadband in rural America has reproduced democratic, social, and economic inequality. There are simply not enough people living in rural America to merit private market investment in broadband. This is why we have public policies to encourage such deployment. Ideally, these policies recognize the important contributions that broadband infrastructure makes to rural America in five distinct ways, which I call the "five pillars of rural broadband": economic development, education, health, civic engagement, and quality of life.

From an economic development perspective, rural small and medium enterprises (SMEs) are particularly harmed by the rural penalty "because they often lack the transaction economies that exist in larger, more metropolitan regions" (Grubesic and Mack 2017, 129). Townsend et al. (2013), for instance, document how broadband in rural Scotland has been a boon to local businesses: "For these practitioners to remain in, and thrive in rural areas, and contribute effectively to their local economies and communities," the authors conclude, "they need to be able to access the tools enabled by adequate broadband" (178). Put bluntly, "Rural communities need good broadband infrastructure if they are to remain viable" (Townsend et al. 2013). The National Rural Electric Cooperative Association (NRECA) estimated that the lack of broadband for electric co-op members "results in more than $68 billion in lost economic value" (Tucker et al. 2018). Meanwhile, a 2016 Hudson Institute study found that rural broadband providers added $24.1 billion to the American economy in 2015 (Kuttner 2016).

Once present in a community, broadband is linked to income growth and slowed unemployment (Whitacre, Gallardo, and Strover 2014). An early study by Peter Stenberg et al. (2009) found that "wage and salary jobs as well as number of proprietors, grew faster in counties with early broadband Internet access" (353). Studies have also demonstrated how the presence (or absence) of broadband is a crucial factor in a business's decision to stay in a rural area (Townsend, Wallace, and Fairhurst 2015) or to relocate *to* a rural community (Kim and Orazem 2017; Lawless and Gore 1999). Strikingly, Younjun Kim and Peter Orazem (2017) found that "rural firms are 60% to 101% more likely to locate in ZIP codes with broadband availability" (286). Broadband is also facilitating a new generation of agriculture ("precision agriculture") and is essential for all aspects of contemporary agriculture, from planting to harvesting to distribution (Gallardo 2016; Whitacre, Mark, and Griffin 2014; see also chapter 4).

Recent studies have also concentrated on the value that high-speed broadband brings to rural housing markets. In a first-of-its-kind study, Steven Deller and Whitacre (2019) convincingly document how "higher access to broadband . . . has a positive impact on remote rural housing values" (15). The authors found the higher the broadband speed, the higher the value increase until speeds over 100 Mbps are reached. Deller and Whitacre's pathbreaking study echoed results from earlier hedonic studies of broadband and housing values. Gabor Molnar, Scott Savage, and Douglas Sicker (2013) found that FTTH increased housing values between 3 and 4 percent, while Russell Kashian and Jose Zenteno (2014) found a 2.7 percent bump in the value of internet-connected houses in a rural Wisconsin community (both cited in Deller and Whitacre 2019). The FTTH Council reports that having access to fiber broadband increases home value by 3.1 percent (see Satterwhite 2015).

Rural education and telehealth also benefit tremendously from high-performance broadband.[15] Rural America is older and in poorer health than the rest of the country but suffers from a lack of doctors, nurses, and adequate insurance (American Hospital Association 2016; Donna Harvey 2019). Telehealth, including remote patient monitoring, record keeping, and e-care, is essential in rural communities but requires a robust and symmetric high-speed broadband connection (2010). As Gallardo (2016) writes, "Because of its detachment from location, telehealth makes it possible to have access to doctors and specialists even in rural clinic or hospital with only a part-time nurse on staff" (chap. 3, sec. 4). Brittney Bauerly et al. (2019) call broadband a "super-determinant," lamenting that "despite telehealth's great potential to improve healthcare access, the promise of telehealth is stymied by the lack of reliable broadband coverage" (40). Adding to this point, Whitacre, Denna Wheeler, and Chad Landgraf (2016) empirically demonstrate how rural health care centers "connect with lower speeds than do their more urban counterparts" (284). This is particularly worrisome for elderly, low-income, and minority patients (Fortney et al. 2011).

Similarly, education, both institutional (K-12, postsecondary) and continuing (including digital literacy), require broadband connectivity (Townsend et al. 2013). Upward of 70 percent of teachers assign homework online, but according to the Pew Research Center, 15 percent of US households with school-age children lack broadband access and 17 percent of teens "say they are often or sometimes unable to complete homework assignments because they do not have reliable access to a computer or internet connection" (Anderson and Perrin 2018). In rural communities, 18 percent of 5–17-year-olds lack broadband (compared with 7 percent of suburban students) (Berdik 2018). This lack of connectivity defines what FCC commissioner Jessica Rosenworcel (2015) calls the "new homework gap," which she describes as the "cruelest part of the digital divide" (2). In rural areas of developing nations, this is particularly acute. Klein (2013) describes how the deployment of telecenters in rural Chile may materially improve community development aims around gender and education. Back in the United States, some rural communities are experimenting with Wi-Fi-enabled school buses to ease the pressure on unconnected students (Bentley 2018).

During the COVID-19 pandemic, the homework gap has only worsened. Parents are forced to drive their students to the parking lots of libraries and McDonald's restaurants to piggyback off of extended Wi-Fi signals (Kang 2020). Similarly, communities are driving Wi-Fi-enabled school buses to unconnected neighborhoods and constructing solar-powered hotspots in others. Meanwhile, schools and libraries are loaning out hotspots faster than they can get them in (Higgins 2020; Khazan 2020). And this is not only a rural issue. A 2020 report by the State Council of Higher Education for Virginia found that "40 percent of all students without access live in or around Virginia's cities" (McKenzie 2020). For both education and telehealth in a time of COVID, broadband has never been more important, or more absent.

A small corpus of studies has also documented how broadband facilitates civic engagement. In an early general population study, Karen Mossberger, Caroline Tolbert, and Ramona McNeal (2012) found convincing evidence of the relationship between broadband use and civic engagement, concluding that access to the internet, in concert with digital skills, facilitates what they call "digital citizenship" (136). For the delineated topic of *rural* broadband and civic engagement, Whitacre and Jacob Manlove (2016) noted that broadband *adoption* (rather than access) was a strong predictor of certain elements of civic engagement like expressing opinion and the likelihood of contacting a local official, but they also found a disturbing negative correlation between broadband adoption and the likelihood of voting. In a later study, Whitacre (2017) demonstrated that a fixed broadband connection is "more strongly associated with taking local civic action than is having a mobile-only connection" (755). This is particularly notable for rural communities since some telecommunications companies are abandoning their fixed connections and moving customers to mobile-only, as noted above (Brodkin 2017b). A series of studies from the United Kingdom have also demonstrated how broadband connectivity contributes to the resiliency of rural communities by strengthening "local rural identity" (Ashmore et al. 2017; Roberts et al. 2017).

What binds these studies together is the belief that the presence of broadband augments quality of life (Whitacre, Gallardo, and Strover 2014; Whitacre and Manlove 2016). As was recounted to me by Bernadine Joselyn, director of public policy and engagement for Minnesota's Blandin

Foundation, "Everything is better with better broadband" (personal communication, 1/15/19). This argument is also not limited to rural communities.
Sharon Strover (2019) documents how access to broadband connectivity via
library-loaned hotspots contributes to a sense of "digital dignity" in New York,
reinforcing a feeling of being "like everyone else" (200). A similar sentiment
was expressed in Duarte's (2017) study of Indigenous broadband networks,
where digital connectivity may serve "Native peoples in their efforts at cultural revitalization and the strengthening of tribal governance" (139).

The presence of broadband—captured not just by availability but also by
affordability and accessibility in terms of digital skills, which together form
the concept of "digital inclusion"—has tremendous impact on rural life. Its
lack has understandably captured the attention of both lawmakers and the
press, to say nothing of those living in "broadband deserts." Yet the fact
remains that despite billions of dollars of subsidy, the rural-urban digital
divide is growing rather than shrinking. As Lorna Philip, Caitlin Cottrill,
and John Farrington (2015) observe, "Internet applications and services are
growing rapidly in their demand on connection capacities, and pressures
from Big Data and, relatedly, the Internet of Things, will leave rural broadband connection speeds trailing in their wake" (168). *Farm Fresh Broadband*
investigates why this gap persists, and what we can do to close it.

The Rural

Thus far, I have employed the term *rural* without much reflection or definition. It's now time for that to change. "Rural"—like the words "local" and
"community"—is a word that we take for granted. It is commonly assumed
to be synonymous with "farm": the fields of Nebraska and Kansas are rural;
the city streets of Manhattan are not. The rural, however, is infinitely more
complicated, contextually, conceptually, legally, and politically (Ali 2018;
Kellogg 2001).

The rural is often taken to mean what it is not: urban (Ilbery 1998;
Thomas et al. 2013). But deciding what is urban is an equally difficult problem. In practical terms, the US Census Bureau (2017) defines as "rural" any
population cluster that is neither an urbanized area (50,000 or more people) or urban cluster (2,500–50,000 people). Urban areas must also "have a
population density of at least 1,000 people per square mile," whereas urban

clusters are "the surrounding census blocks that have an overall density of
at least 500 people per square mile" (2011a).

Geography also has much to do with this definition of rural and urban.
Should a "rural area" be defined as a population center (census), a land-use
center ("urban area") or an economic center (micro/metro)?

> Depending on the boundary choice and the population threshold, the share of
> the U.S. population defined as rural and its socioeconomic characteristics vary
> substantially. In 2000, 21 percent of the U.S. population was designated rural
> using the Census Bureau's land-use definition (outside urban areas of 2,500 or
> more people), compared with 17 percent for economically based nonmetro areas
> (outside metro areas of 50,000 or more). (Cromartie and Bucholtz 2008)

These disagreements make it difficult for lawmakers and regulators to
enact policies targeted toward rural areas. In 2008, the USDA was ordered
by the Farm Bill to categorize how it defines "rural" (Food, Conservation,
and Energy Act of 2008, §6108). The department's 2013 report found more
than thirty different definitions in use (USDA 2013). Its Telecommunications Loan and Grants program (see chapter 3), for instance, uses a population threshold of 5,000, while its new Broadband ReConnect Program
has a threshold of 20,000. The FCC's definition is equally inchoate and
inconsistent, changing with the subject of its inquiry. Set in 2004, Section
54.600(b)(1) of the Commission's rules defines rural for the Rural Health
Care program as an area entirely outside of an urban area with a population
of 25,000 or greater (see FCC 2014b, fn. 13). In an inquiry into television
frequencies, however, the FCC (2014a) defined rural as any area where there
are multiple unused television channels, while in another spectrum-related
inquiry, "rural" is defined as "a county (or equivalent) with a population
density of 100 persons per square mile or less, based upon the most recently
available Census data" (FCC 2012).

To get us started, I employ the definition found in the 2008 Farm Bill.
This bill clarified the meanings of "rural" and "rural area" and allowed the
secretary of agriculture some flexibility in categorizing areas that are "rural
in character":

> IN GENERAL.—Subject to subparagraphs (B) through (G), the terms 'rural' and
> 'rural area' mean any area other than—
>
> (i) a city or town that has a population of greater than 50,000 inhabitants;
>
>   And (ii) any urbanized area contiguous and adjacent to a city or town described
>   in clause (i).

AREAS RURAL IN CHARACTER.—

(i) APPLICATION.—This subparagraph applies to—(I) an urbanized area described in subparagraphs (A)(ii) and (F) that—(aa) has 2 points on its boundary that are at least 40 miles apart; and (bb) is not contiguous or adjacent to a city or town that has a population of greater than 150,000 inhabitants or an urbanized area of such city or town; and

(ii) an area within an urbanized area described in subparagraphs (A)(ii) and (F) that is within 1/4-mile of a rural area described in subparagraph (A). (Food, Conservation, and Energy Act of 2008, §6108)

This definition is useful in its simplicity and clarity, and it focuses on population, density, and geography. It is also the basic definition used by Gallardo (2016) in his book *Responsive Countryside*. Using this definition, the Census Bureau notes that 60 million people live in so-called rural America. Rural America makes up 97 percent of the country's landmass but only 19.3 percent of the population.[16] This population is also shrinking. According to USDA's *Rural America at a Glance, 2017 Edition*, between 2000 and 2016, rural America lost 200,000 residents—its first net population decline in recorded history. This report also notes that rural America tends to be economically disenfranchised compared to urban America, both in terms of employment growth (which had a growth rate of only 0.4 percent compared to 1.5 percent in urban America) and poverty (16.4 percent vs. 12.9 percent in urban America). Gallardo (2016) also reports that rural America is older than urban America and is getting older because of out-migration of young people (in 2014, 42 percent were older than forty-five). The hardest-hit areas in this population and economic decline were those dependent on agriculture in the Great Plains, Midwest, and southern Coastal Plains. The Great Plains and Midwest represent the geographic areas of my research, making this decline particularly notable.

Another set of definitions moves us away from geography and demography toward the conceptual and postmodern (Halfacree 1993; Ilbery 1998; Phillips 1998). We may think of these either as "rural as a locality," suggesting that "if rural localities are to be studied in their own right, they must be carefully defined according to those characteristics which make them rural," or "rural as a social-representation," whereby rurality is defined by discourse (Halfacree 1993; Ilbery 1998, 3). In this tradition, many suggest the search for a singular definition "is neither desirable or feasible" (Halfacree 1995, qtd in Ilbery 1998, 3), all the while recognizing that it "remains an important category because behaviour and decision making are influenced by people's perceptions of rural" (Ilbery 1998, 3). In this capacity, the rural may be what poststructuralists call a "floating signifier" or what Christina Dunbar-Hestor (2013) calls a "discursive boundary object" with loose definitional perimeters.

These floating definitional boundaries (intentional or otherwise) make it difficult to theorize and research the rural—a task already woefully under-complete, even without these complications (Friedland 2002; Ilbery 1998a; Thomas et al. 2013). Most often in conversations of social theory the rural is taken to be a porous term—whatever scholars and rural residents choose it to be (Halfacree 1993; Ilbery 1998). If not, then critical theorists tend to dismiss the rural as uninteresting (e.g., Marx [1939] 1993), perceive it as something that needs fixing (Berry 2006; Oldenburg 1999), or point out our propensity to fetishize the rural through nostalgia (Thomas et al. 2013; Williams 1975). Karl Marx ([1939] 1993) paid little heed to the rural aside from seeing it as a feeding ground for the urban workforce. In the *Grundrisse*, he wrote of the "idiocy of rural life" and "the detrimental effects of both the rural oppression from which he saw Capitalism as a great savior and the urban (Capitalist) oppression from which socialism would save the masses" (Thomas et al. 2013, 37). Friedrich Engels (2009), in *The Condition of the Working Class in England in 1844*, was slightly more forgiving. For his part, Henri Lefebvre (1995, 2016a, 2016b), after cutting his sociological teeth on treatises about the rural (including his doctoral dissertation), abandoned it in favor of more urban pastures (see Lefebvre and Gaviria 1973). Raymond Williams (1973) gave more consideration to the cultural differences between the country and the city, pointing out our propensity to romanticize a pastoral life that never existed. To fetishize the rural this way does disservice to those who live there and fails to acknowledge what Michael Bell et al. (2010) call "the power of rural voices": "As scholars, we need to articulate the active voice of the rural in order to understand its constant articulation and rearticulation through mobilization and stabilisation, however progressive or deplorable these articulations and rearticulation may be" (220).

In privileging the voices of rural America, two distinctions must be acknowledged. First is the difference between "rural" and "farm." As the W. K. Kellogg Foundation (2001) reported, most respondents to its survey tended to equate rural with farm and "perceive that rural America is serene and beautiful, populated by animals and livestock and landscape covered by trees and family farms" (1). While I use case studies and examples from the agricultural communities of the American Midwest, critical rural studies teaches us not to conflate "rural" with "farm," for while agricultural communities exist within the rural, not all rural communities are agricultural in nature (in fact, very few are) (Friedland 2002; Thomas et al. 2013). The second acknowledgment is to disassociate the notion of "rural" from the presumption of "whiteness" (Woods 2005). The presumption of whiteness and rural America assumes a homogeneous experience with the digital divide, a presumption debunked with the aforementioned discussion of the intersectionality of digital distress. As Michael Woods (2005) explains, the assumption of rural America as a "white space" excludes people of color "from discursive representations of the rural idyll" and negates their experiences of "racist discrimination in their occupancy or use of rural space" (282). Not only does it erase the experiences of people of color who have lived in rural America for generations, but it sees immigration as threatening and intrusive: "ethnicity is seen as being 'out of place' in the countryside, reflecting the Otherness of people of colour" (Agyeman and Soppner 1997, 199). It also fails to recognize the fact that rural America is becoming more diverse (Brown 2014; K. Johnson 2014). The Pew Research Center found that the white share of rural America is shrinking, now representing 79 percent of the rural population (K. Parker et al. 2018). Immigrants were responsible for 37 percent of the population growth in rural counties, while minorities were responsible for a whopping 83 percent of the population growth of rural America from 2000 to 2010 (K. Johnson 2014).

Rural America is having a moment in the US popular *and* political consciousness, which makes my research and the topic of rural broadband especially timely. The election of Donald Trump as president in 2016 focused national attention on the so-called left-behind places neglected in the industrial swing to digitalization and the collapse of manufacturing, natural resource extraction, and the move to industrial agriculture (Hendrickson et al. 2015). Broadband is a key component in keeping people living, working, and thriving in rural America and has seen its celebrity rise alongside that of rural economic development. There is no shortage of ideas for how to connect rural America, illustrated poignantly in the twenty bills on the topic proposed in Congress in 2018. In 2020, COVID-19 cemented rural broadband to the forefront of social and political conversations in the country, making this an opportune moment for analysis, critique, and reflection.

This book joins these conversations by unpacking the policies that govern the deployment of rural broadband, the politics that keep deployment from occurring, and the stakeholders invested in its success and failure.

Methodology: Lived Policy

The primary research question driving this book is about money, or, more conceptually, political economy. Specifically, I ask: Where is the (approximately) $6 billion a year the federal government spends on rural broadband subsidies going? To this, I add a question about experience: With a particular focus on agricultural communities, how is broadband policy lived and experienced in rural America? I relied on an interconnected set of methods and theories to address these questions. First and foremost was an analysis of thousands of pages of policy documents. Sources ranged from the FCC and RUS to public comments filed to these agencies, to the various Farm Bills and congressional proposals, and reports from the CRS and Government Accountability Office (GAO). Those familiar with the policies of rural broadband will find a dedicated engagement with the FCC's 2009 Rural Broadband Strategy, the Broadband Opportunity Council's 2015 investigation into rural broadband, the USDA's 2018 ReConnect Program, and the FCC's 2018 Rural Digital Opportunity Fund (RDOF). Each of these portfolios received over a thousand pages of public comments, and I read and analyzed every page.

Accompanying this policy analysis is a series of in-depth interviews with key stakeholders in rural broadband policy and deployment. Like with my previous research projects (Ali 2017a; Ali et al. 2018), I employed open-ended and nondirective questions, which allows for more storytelling and resembles a "natural" conversation (Atkinson 2007; Lewis 1991). Interviews were conducted with a variety of stakeholders: informants at the FCC and RUS; rural broadband researchers and advocates; industry associations like the NTCA— The Rural Broadband Association and NRECA; and state representatives, journalists, consultants, rural broadband providers, rural residents, and farmers.

Central to my interview selection is the mix of respondents. This book is based on interviews not only with elites, as commonly found in policy research (including my own previous work), but rather with a variety of stakeholders. In total, I conducted ninety interviews, with sessions lasting between thirty minutes and four hours. The need for variety in respondents was apparent in the topic: rural broadband is not a policy issue; it is a lived issue. Calls for a critical approach to the rural require us to privilege its many voices. As Bell et al. (2010) write, "We can and do hear the active rural voice because it is a voice of power" (213). This requires a distinct methodological approach beyond that of standard policy analysis and elite interviews. I think about this as "lived policy," which describes how public policies are articulated and experienced in everyday life.

I draw here from important work done in theology on "lived theology" and "contextual theology," which is based on connecting theological studies with "the historical, socio-cultural, political and cultural realities" of the people who live and practice religion (Bevans 2002; C. Marsh, Slade, and Azaransky 2017; Matheny 2012, xi). Just as "theological education should never lose sight of the problems and realities of the churches and people that they serve" (Matheny 2012, xii), I argue that public policy and the critical study of public policy cannot forget the realities of the people they strive to serve. Rural broadband is lived hundreds of miles away from the hearing rooms of Washington, DC, and the corporate offices of AT&T and Verizon. It is experienced (or not) in towns and communities across the United States, and it is the dialogue between the articulation of policy and the experience of policy that I endeavor to understand.

To embody this method of lived policy, I undertook a series of field visits, participant observations, and interviews. In 2015 Christian Herzog and I wrote about the need for critical policy scholars to incorporate more ethnographic methods and reflexivity into their research design (Herzog and Ali 2015). This book is my opportunity to follow through on this call to action. I began by attending a number of industry events, conferences, and webinars as a form of participant observation to get a sense of how the broadband industry was discussing rural broadband. I then began conducting in-depth interviews by phone. Next, for two weeks in the summer of 2018, my hound dog, Tuna, and I drove across the American Midwest in what I called our "rural broadband road trip." Driving from Charlottesville, Virginia, to Winnipeg, Manitoba (Canada), via Kentucky, Indiana, Missouri, Iowa, Minnesota, and North Dakota, Tuna and I covered over 3,600 miles to learn firsthand about rural broadband. I spoke with residents in small communities; rural broadband advocates such as the civil society group Connected Nation, state representatives such as Tim Arbeiter who, when we met, was just two weeks into his new job as the director of Missouri's Broadband Development Office; town and county officials such as Kyle Oldre, county administrator of Rock County, Minnesota; and rural broadband providers in Minnesota, Iowa, and South Dakota. While I do not claim that this two-week trip is a true ethnography, which takes thousands of hours of immersion, it nonetheless employed ethnographic methods like field visits, interviews, and participant observation (Hammersley and Atkinson 1995) to better understand how broadband policy is lived in rural America. I chose the Midwest (which, according to the Census Bureau, includes Illinois, Indiana, Iowa, Kansas, Michigan, Minnesota, Missouri, Nebraska, North Dakota, Ohio, South Dakota, and Wisconsin) to ground this study for five reasons. First, this region has experienced the bulk of out-migration in rural America. Second, the region falls well below the national average of broadband connectivity. Third, it receives a significant portion of federal broadband funding and houses examples of successful state intervention (Minnesota). Fourth, the Midwest is home to the majority of electrical and telephone cooperatives providing broadband to rural communities. Fifth, the region has been particularly vocal at the federal policy level, exemplified by the volume of comments to FCC inquiries and rulemakings.

All data—document, interview, and ethnographic—were analyzed using a combination of thematic coding analysis and grounded theory (Glaser and Strauss 1967; Herzog, Handke, and Hitters 2019; Hesse-Biber and Leavy 2010). Thematic coding analysis is the search for patterns in a series of texts. These patterns—which may enter the dozens—are then refined into fewer, more abstract themes. These themes become the basis for the argument and reporting. This is akin to the grounded theory method of Barney Glaser and Anselm Strauss (1967), where data is mined for "incidents" that are then assembled into patterns and themes, with the ultimate goal being theory generation. The end result is robust hermeneutic analysis, thick description, and, through this, the development of what Joe Kincheloe and Peter McLaren (1994) call "critical trustworthiness" of data interpretation.

Based on these methods, I argue that rural broadband policy is both broken and incomplete. It is broken because it lacks coordinated federal leadership and is mired in a battle between regulators, a bewildering array of rules, inconsistent state intervention, the dominance of large providers, and a policy process that favors these companies over the hundreds of local ISPs. It is incomplete because it fails to recognize the crucial role of communities, cooperatives, and local ISPs in providing broadband to rural people. To be sure, the $6 billion allocated for rural broadband deployment is being spent on rural broadband, but it is not being spent efficiently or democrati-cally. Here, the political economy of rural telecommunications mirrors that of industrial agriculture: a sector dominated by massive corporations and a regulatory system designed to favor them. Take AT&T, CenturyLink, and Frontier, for example. These incumbent telecommunications companies receive hundreds of millions of dollars a year to connect rural America, but they refuse to upgrade their networks, arguing the 1990s-era technolo-gies of DSL and copper wires are good enough for rural connectivity. Fur-thermore, they parlay their incumbent status to consume the majority of federal subsidies for rural broadband, to the detriment of small ISPs and cooperatives. On top of this, the FCC and RUS—the two primary broad-band regulators and subsidizers—refuse to share information and cooper-ate, thus leading to knowledge silos and inefficiencies that further protect incumbent providers.

To rectify these inequalities, I argue that lawmakers (both legislative and administrative) need to democratize the policy architecture of rural broad-band. This means recovering the inertia at the federal policy level that brought about the wiring of rural America for electricity in the 1930s and telephony in the 1950s. This means strong federal leadership, data sharing, and comprehensive mapping. As I argue in chapter 3, it means recogniz-ing the potential of RUS and harnessing its capacities to lead this charge, rather than the contested policy field currently occupied by the FCC, RUS, and NTIA. It also means equalizing the federal subsidy system. Currently, the FCC and RUS control the bulk of rural broadband subsidies. As I report in chapters 2 and 3, their policies favor incumbent, legacy telephone com-panies, dissuade new entrants and competitors, and put onerous condi-tions on small providers. The ten largest telecommunications companies, for instance, received over $1 billion a year from the Universal Service Fund (USF) from 2015 to 2020, and only had to build networks to a speed of 10/1—speeds that have been obsolete for years. In contrast, hundreds of firms must split the other $1.2 billion, and have much higher speed thresh-olds. Imagine if the entirety of these billions could be distributed to orga-nizations that bid and promise to connect their communities, rather than being earmarked for the largest providers and technologies?

In democratizing rural broadband policy and subsidy, policymakers also need to recognize the important role cooperatives play in the rural broad-band ecosystem. More than this, cooperatives need to be actively supported at the federal level through increased subsidy, and at the state level by legislation that promotes their deployment of broadband. Said differently, I argue that cooperatives are key stakeholders in rural broadband. As I note in chapter 1, electricity and telephone cooperatives were essential to the wiring of rural America in the 1930s, 1940s, and 1950s. Today, they are crucial alternatives to the major telecommunications companies. Unlike major telecommunications companies, cooperatives are not driven by shareholder dividends and short-term returns on investments that dissuade them from servicing rural areas. Instead, they are committed to community improvement and to their community-based members. Their efforts, however, have been hampered by cumbersome state legislation, opaque funding regulations, inadequate grants, and hostility from incumbent providers (and even other co-ops!).

In sum, democratizing rural broadband policy means changing the way rural broadband is funded, adopting ambitious definitions, mapping broadband deserts, and, most importantly, recognizing that the unsung heroes of rural broadband are the local municipalities, cooperatives, and digital champions who work without recognition to bring connectivity to their communities.

A Brief Note on Theory
This book is firmly rooted in the core tenets of critical political economy of communication (CPEC), in the tradition of Vincent Mosco, Leslie Shade, Robin Mansell, Robert McChesney, Victor Pickard, and Dwayne Winseck as I am literally following the money.[17] CPEC is predicated on the critical interpretation of power and control in the distribution of resources. As Vincent Mosco (2009) defines it, "Political economy is the study of the social relations, particularly the power relations, that mutually constitute the production, distribution, and consumption of resources, including communication resources" (2). From our perspective, it is not too much of a stretch to think of broadband as a resource, a utility, and a necessity for contemporary existence. Thinking this way allows us to interrogate why it is that broadband is not near-universal in the United States, as are water and electricity. Equally, it allows us to determine what structures and power dynamics stymie broadband's universality. Lastly, it allows us to put rural broadband policy in a historical context and within a wider social totality. This means we can critically compare the federal approach to broadband against similar approaches to "wire" rural America with electricity and telephony in the

1930s and 1940s. What makes CPEC distinctly critical is its push to actively intervene in the power structures it assesses and the status quo claims it challenges. This is known as praxis: the union of research and action toward the goal of human emancipation (Ali and Herzog 2018; Habermas 1985; Marx 1888). *Farm Fresh Broadband* embodies this belief and concludes with recommendations for rural broadband policymakers and practitioners to achieve the desired goals of universal availability and adoption.

Central to CPEC, critical communication policy studies, and my book is the articulation of power and legitimacy (Braman 2009; Freedman 2014; Mansell 2012; Mosco 2009).[18] As Steven Lukes (2005) reminds us, power has three faces: (1) *A* getting *B* to do something *B* would not normally do; (2) *A* setting the agenda for what is legitimate for *B* to consider; (3) *A* shapes *B*'s preferences in a way that appears natural. In order for certain types of power to be articulated and maintained, legitimacy is necessary. Legitimacy "is a generalized perception or assumption that the actions of an entity are desirable, proper, or appropriate within some socially constructed system of norms, values, beliefs, and definitions" (Suchman 1995, 574). In the coming chapters, we will see the desire of the power elite in American broadband to maintain their legitimacy while at the same time, unconventional actors, like the RUS, seek (and often fail) to create legitimacy in the regulatory space. Part of what *Farm Fresh Broadband* assesses, therefore, is how various stakeholders in rural broadband policy make claims to legitimacy and how these claims are often leveraged to maintain their own monopoly power.

Where I expand on CPEC, communication studies approach to power, and critical communication policy is in my engagement with critical theories of the rural (see Ali 2018). Critics in this vein have pushed us to take seriously the experiences of rural life; to abandon notions of a bucolic, pastoral, singular, and nostalgic rural existence; and to avoid "urbanormativity"— "the general view of urban as normal and real, and rural as abnormal and unreal, or deviant" (Thomas et al. 2013, 5). This has led scholars outside of communication studies to call for a "critical approach to the study of rural areas" (Thomas et al. 2012), a "critical rural theory" (Phillips 1998), or, as Darin Barney (2011) calls it within the field of communication studies, "critical *agri*cultural studies." I myself have called for a distinctly "critical theory of rural communication" (Ali 2018). Particularly important here is the need to privilege rural voices and to recognize the power therein (Bell et al. 2010). This fundamental belief connects the critical theoretical tone of this book with the practical "lived policy" methodological toolkit I developed and described above.

The Book

The scope of the book gradually narrows, moving from a broad historical overview in chapter 1 to policy analysis in chapters 2 and 3 to case studies in chapters 4 and 5. Constructed in such a way, these chapters can be read as standalone segments or as part of the larger narrative.

True to the tenets of political economy, chapter 1 begins by placing rural broadband in historical context through a description of how rural America was wired for electricity and telephony in the 1930s and 1940s. I ground this description in the concept of market failure, arguing this concept defines the history of wiring rural America. It defines the present wiring of rural America as well. I argue strong federal intervention has always been necessary to connect rural America because the competitive market system is anathema to universal service. In the second part of the chapter, I draw on evidence from my interviews to connect the past and the present by comparing rural broadband and rural electrification. This supports the argument for strong federal regulatory intervention in rural broadband.

Chapter 2 moves from past to present and introduces the reader to the policies of rural broadband undertaken by the FCC. Profit and incumbency define these policies, culminating in two interconnected results. First, rural broadband policy represents what Pickard (2015) calls a "policy failure"—defined as a failure to act in the wake of market failure. Second, rural broadband policy is dominated by a "politics of good enough" that serves the material interests of telecommunications companies. Said differently, substandard broadband requirements are justified using the language of "good enough," a discourse repeated by the largest providers. The result of this "discursive capture" (Pickard 2015) is inadequate rural connections. This chapter thus asks, what happens when "good" is the enemy of great?

Chapter 3 continues my review of rural broadband policies but shifts the focus from the FCC to the USDA. Here, I explore the legitimacy and authority of RUS to enact broadband policy, arguing that it represents a "reluctant regulator." In other words, RUS has the potential and legitimacy to be a major actor in the rural broadband policy arena but fails to actualize this position for a variety of political and economic reasons.

Chapter 4 embodies the method of lived policy and moves the policy conversation outside Washington, DC, and into the states, counties, and cooperatives of the rural Midwest. This chapter tells the story of Rock County, Minnesota, and its success in bringing fiber to 99.93 percent of residents through an innovative partnership with Alliance Communications, a telephone cooperative located in neighboring South Dakota. I argue here that rural broadband is a fundamentally local phenomenon, and federal policy needs to recognize the role played by local ISPs and cooperatives. Rural broadband policy and deployment is not solely the purview of the federal government and major telecommunications companies. In fact, it requires just the opposite. It demands a multistakeholder approach that is currently not recognized within the status quo of DC politics.

Chapter 5 focuses attention on an application of broadband in rural America, namely precision agriculture. "Precision agriculture" describes the application of information and communication technologies to agriculture, from seeding to harvesting to distribution. At present, broadband infrastructure—both wireless and wired—is incapable of supporting a fully actualized precision agriculture system. As a result, politicians and industry organizations jump at this connection, using precision agriculture as an economic rationalization for rural broadband deployment. While important, I argue this exuberant discourse of rural broadband and precision agriculture masks deeply entrenched and unresolved political economic issues that span both telecommunications and agriculture industries. This chapter extends the conversation on rural broadband applications by introducing a host of tangential issues that come with a connected farm in the form of data ownership and privacy. Policymakers and industry watchers need to be more critical of these practices and discourses. Until these issues are resolved, a more temperate discourse is required so as not to cede rural Americans' digital sovereignty to the major agribusiness and telecommunications companies and further exclude and disenfranchise the farmer and rural broadband consumer.

The conclusion is my opportunity for praxis. Here, I lay out my vision for rural broadband policy in the United States, ultimately arguing for a national rural broadband plan. My proposed national rural broadband plan would appoint a lead agency for rural broadband, reform subsidies, encourage data sharing between the FCC and RUS, require granular broadband data collection, and set aside substantial funding for skill development and digital inclusion. I do not argue for nationalized broadband, but rather for a multistakeholder broadband system guided by insightful public policy and funded through numerous democratized revenue streams, including industry cross-subsidy, public funding, and private, market-based support. A national rural broadband plan would show that the United States is serious about both global competition and quality of life of rural communities. As it stands today, a lack of universal broadband means rural communities are suffering. We are failing to ameliorate these conditions because we are not taking all stakeholders into account. We are failing because of a lack of coordinated and coherent policies. We are failing because major telecommunications companies get the bulk of funding and fail to deliver. We are failing because the agencies in charge of rural broadband do not even know who has broadband and who does not. A renewed federal commitment to rural broadband would encourage states to take on greater coordination roles, which would then empower local communities to make informed choices about their connectivity. Rural broadband will not be solved by incumbents, regulators, or states. It will be solved by local communities and local companies. That said, robust, coherent, and comprehensive policies can make their efforts so much easier. As I wrote in a recent op-ed in the *New York Times*, "It's time regulators gave these communities a hand" (Ali 2019c).

Bran Flakes to the Home

During my rural broadband road trip, one provider recalled a story of trying to convince his community to adopt FTTH. In response, someone said, "You mean like bran flakes at my house every morning? What do you mean fiber to my home?" In today's connected world, broadband is as essential as bran flakes—one item in a nutritional toolkit of rural community development. Current policies do not encourage us to eat our digital fiber. Instead, money, markets, politics, and profits will define the forthcoming pages in this book. We will delve deep into the market failures of rural communication; we will learn how federal policies prioritize large telecommunications companies at the expense of small ISPs and co-ops; and we will see how many of these companies deliver only the legally required minimum speeds rather than treat these minimums as a floor to build on.

As much as rural broadband policy is a story of failure, it is also one of ingenuity and innovation in America's heartland. It is here where we find some of the successes of rural broadband—like Luverne, Minnesota, seat of Rock County, whose leaders risked a $1 million bond to bring fiber-optic broadband to the county, or the PRTC, which brought fiber-optic broadband to McKee, Kentucky, one of the poorest communities in the state. These are the local companies and cooperatives more interested in serving their members and communities with a public service than earning a short-term return on investment. And there are many who are in need of these providers, like farmers and growers, who are all too often left out of the conversation, but for whom broadband to the farm would mean a new era of agriculture.

The story of rural broadband in the United States is equally one of failure and one of promise. *Farm Fresh Broadband* captures both, all the while reminding the reader that as much as technology policy is enacted in Washington, DC, it is lived throughout the country. Rural broadband policy, like all public policy, is a living, breathing, and changing creature, which means that, with the right attention and coaxing, we can make it work for the people who need it most.

1 The Struggle for a Wired America: A Brief History of Connecting the Countryside

I was sitting in my home office in Charlottesville, Virginia, on an unseasonably warm Valentine's Day in 2018 talking to Roger Johnson and Matt Perdue of the National Farmers Union (NFU). My conversation with Johnson and Perdue was not about weather, however, but about the role of the NFU in promoting rural broadband deployment. Like its distant cousins the AFBF and the National Grange of the Order of Patrons of Husbandry ("the Grange"), the NFU has been championing rural broadband for years, and it included access to competitively priced, high-speed broadband in both its 2017 and 2019 national policy priorities. During our conversation, I asked Johnson, president of the NFU, why broadband was important for rural communities. His answer tied the contemporary discussion of rural broadband with a much older conversation: electricity for all.

> There's a couple reasons why it matters. There's obviously the connectivity issue that is not unlike the same issue we faced generations ago when telephones came to the countryside, and when electricity came to the countryside, which is to say that in neither case did either of those services come to the countryside until they had been for some while in the more concentrated urban areas. And they ended up only coming to the countryside because of a focused public policy kind of effort that encouraged investment and facilitated folks in the countryside, themselves, organizing and providing services predominantly through the use of cooperatives. (personal communication, 2/14/18)

This nostalgic connection to electricity is a sentiment I heard repeatedly in my conversations with rural broadband stakeholders, and one often repeated in public comments to the FCC and RUS. As Perdue, government relations representative for the NFU, explained further:

> So, you know Roger [Johnson] referred to Rural Electrification Act. We've known since the 1930s that infrastructure like this is not going to get you the same return

on investment as a similar investment would in an urban area. And I think that remains the biggest obstacle. It's very difficult for a private company to pay to create and install the infrastructure that is only going to get them to a few homes, when that same amount of infrastructure could connect an entire square mile of a large city. (personal communication, 2/14/18)

What both Johnson and Perdue are referring to is the New Deal effort to bring electricity to rural America through an ambitious agency called the Rural Electrification Administration (REA).[1] Created in 1935, the REA was one of two essential actors in the electrification of the country. Its role was to encourage the creation of electric cooperatives and subsidize the distribution of electricity to co-op members through self-liquidating loans. This proved tremendously successful, begetting over 900 electric co-ops and bringing electricity to millions of Americans. In 1949, rural telephony came under the auspices of the REA, with the same drive and zeal. The end result was that by 1960, twenty-five years after the creation of the REA and only eleven years after the creation of the telephony program, nine out of ten farmhouses were electrified and two-thirds had telephony.

The second actor in the history of connecting the countryside is the rural communities themselves, manifest through the hundreds of electrical and then telephone cooperatives they formed. With grassroots, local organization, coupled with a comprehensive federal policy apparatus and dedicated federal funding, rural communities that private companies deemed unsuitable for service were wired and connected.

The parallels with rural broadband, some eighty-five years after the launch of the REA, are apparent and powerful. Private industry refused to serve rural America because of the lack of return on investment, which led to market failure first in electricity, then in telephony and today in broadband. For those lucky few who had service, they had to pay for the installation of the lines themselves and even then did not own them—not unlike the example of the Schneider farm that began this book. What is worse, rural households paid a tremendous amount for electricity, often twice the standard urban rate (Brown 1980). Lack of availability and high rates also define rural broadband. On the flip side, communities, cooperatives, and municipalities organized and connected themselves when the bastions of capitalism abandoned them. This, again, is true today. Finally, none of this would have been possible without a dedicated federal agency with the resources and authority to get it done. Indeed, economist John Malone (2001) has hailed the REA as "one of the most immediate and profound successes in the history of federal policy-making" (5).

With a dual focus on the REA at the top and local communities at the grassroots, this chapter recalls the significant moments in the structural history of connecting rural America. As its subtitle suggests, this is a *brief* history; I cannot possibly do justice to the complex and poignant history of both rural electricity and telephony in a single book chapter. Entire volumes have been written on the matter.[2] Instead, the point of this chapter is to highlight key moments in the structural (rather than cultural) history of rural connectivity and, in the final section, connect this history to the contemporary experiences of my interviewees regarding broadband. Doing so reveals dynamic similarities and lamentable contrasts, for the fact remains that the United States was far more effective at connecting its citizens with electricity and telephony than it has been with broadband. I argue in this chapter that just as a strong federal presence was necessary to connect rural America with those utilities then, the same is required today for broadband. I begin with an overview of rural electricity and telephony and conclude by linking these histories forward to my interviews. Here I draw out the notions of market failure and federal intervention as a key obstacle and pillar, respectively, in the drive for universal connectivity. So strong is this connection between electricity and broadband that one respondent simply stated, "Broadband is the next electricity." This chapter explains why.

Rural Electrification

Large-scale power production began in the United States in the 1880s with Thomas Edison's construction of New York's first central (power) station in 1882. Throughout the tail end of the nineteenth century and in to the early twentieth, the political economy of electricity was linked to industry and business, with domestic power provision more of a necessary evil than an economic driver. Power companies were required to provide household service, but their profits were driven by industry. Indeed, "by 1920 only 47.4 percent of city and rural nonfarm residential dwellings had been electrified" (Hirsh 2018, 301), and these customers generated scant revenue. Historian Ronald C. Tobey (1996) writes in his well-regarded history of electrical modernization, *Technology as Freedom*, "In 1929, domestic customers constituted 82.7 percent of users of central station energy, but returned only 29.4 percent of total revenues. Large and small industries, by contrast, constituted 17.3 percent of customers and returned 55.5 percent of total revenues" (10). Of course, those domestic customers were also those who could afford to pay the high prices necessary to subsidize industry use (Kline 2000). Everyone else was left in the dark, literally.

The rationale for the refusal of power companies to connect rural America was based in economics. Rural communities were simply too sparsely populated and too vast to be wired. Those that were connected paid exorbitant rates per kilowatt hour, sometimes double the amount paid by urban users (Brown 1980). In sum, rural electricity, like rural telephony later in the century and rural broadband today, represented a market failure.

Market failure, as the introduction to this book recounted, defines a condition wherein the private market fails to adequately invest in a necessary good and therefore fails to produce socially desirable outcomes (Bator 1959; Medema 2009; Pickard 2013, 2014a, 2014b; Stiglitz 1989). In this case, the private market was unwilling to connect rural America because of a lack of acceptable return on investment: the distances were too vast and the areas too sparsely populated. This was argued in concert with the belief that rural American households went without electricity because they did not want it, not because they could not afford it (Tobey 1996, 120). Electricity, as we will learn below, was thought to be "inelastic"; that is to say, cost was not a factor in a consumer's decision to connect to the grid.[3] It didn't matter how low the price was, rural Americans simply did not want electricity. Since cost apparently didn't matter, power companies refused to lower their prices. This assumption, of course, is false. Rural Americans desperately wanted electricity, but they wanted it at a fair and reasonable price (Kline 2000).

In addition to the classic market failure explanation for the lack of rural electrification, historian Richard Hirsh (2018) adds that utilities avoided rural America "because they viewed farmers with derision—as backward, unmodern, and unsophisticated" (304). Much like today, utilities felt no social or moral compunction to connect the unconnected. Despite this dismissal by the incumbent power companies, farmers eschewed these companies' urbanormative beliefs and were early adopters of modern technologies, including the radio, telephone, automobile, and yes, even power (Hirsh 2018; Kline 2000).

Neglected by "Big Power," municipalities in the early 1900s entered the electricity business, finding inspiration in the government-owned hydro plants of Ontario, Canada, which began its plan for public power in 1906 (Tobey 1996).

Some municipalities extended their lines into rural areas, including the 1906 construction of a line serving five farms in Oregon, and rural lines in California, Colorado, Indiana, Illinois, and Washington appeared in 1910 (Tobey 1996). These, however, were exceptions to a rule dominated first by industry and second by urban, wealthy communities. Far down the priority list of customers were farmers (Pence 1984). Ironically, they were the ones who needed electricity the most. A 1984 commemorative book for the REA's fiftieth anniversary produced by the NRECA recalls the hardships of farming life without electricity, including the need to manually pump and transport water, the inability to refrigerate milk, the intense heat produced by woodstoves and ovens, the backbreaking work of canning and jamming and the monotony of washing (Pence 1984). Crucially, these tasks involved a substantial female labor force, the much-lamented "farm wife," whose archetype would be a leading symbol for rural electrification in the 1920s and 1930 (Kline 2000).

According to Morris Cooke (1939), the REA's first administrator and an early evangelist and architect of rural electrification, discussions around electricity in agriculture began to gain voice around 1910 but were "hardly to be described as representing any deep-seated conviction or interest" (438). This we can assume was only from industry stakeholders and policymakers, because rural Americans had already begun to experiment with current. Before systematic government intervention and an established policy goal of universal service that originated with the REA, farmers and rural communities had a do-it-yourself (DIY) and distinctly communitarian approach to electrification (Hirsh 2018; Kline 2000). Early attempts to electrify the farm noted by Ronald Kline (2000) include powering turbines by homemade dams, using wind power, and relying on gasoline engines. More formal attempts occurred through the formation of electric-power cooperatives. In an anachronistic twist of fate, the early electric-power co-ops were inspired by the rural telephone cooperatives, which emerged at the start of the century. As Kline (2000) summarizes, however, there are important differences between the two:

> Farmers could build private telephone systems without having to connect them to each other or to the urban world. The low-voltage and low-current equipment was relatively inexpensive, and local people operated the switchboard and ran the co-op. Far more complex technology and expertise were associated with running large engines and dynamos to generate thousands of kilowatts at a central source, stepping it up to several thousand volts, transmitting it safely to farms, then stepping it down to 100 volts for lights and appliances. (100)

Because of these complexities in power generation, many electrical cooperatives bought electricity wholesale from utilities and distributed it to their members. Of this, Hirsh (2018, 297) makes the important contribution that despite the popular belief that rural electrification began and ended with the REA, the Tennessee Valley Authority (TVA), and President Roosevelt, the utility industry and land-grant universities made significant progress in rural electrification in the 1920s, achieving 11 percent electrification of American farms by 1931. Other sources note, however, that only half of these connected farms used utility company power, while the other half used electric-lighting plants (Nye 1990). Obviously, much work still needed to be done.

In the 1920s, the driving progressive discourse in favor of rural electrification was not as much about productivity as about social welfare, most notably the need to "eliminate drudgery, save labor, and increase leisure time" for the rural farm wife (Kline 2000, 105). As Meade Ferguson, chair of the Home Economics Committee of the Grange, said in 1935, "Adequate electric current for farm homes will mean the emancipation of farm women from the endless household drudgeries which they have borne too long" (qtd. in Kline 2000, 3). The farm, the house, the wife, were all part of a larger social engineering project of modernization (Kline 2000; Tobey 1996). "Electrification," writes Kline (2000), "was seen as the measure of farm men and women, tangible evidence of how far they had advanced on the road to modernity" (131). More than the altruistic push to modernity, however, rural electrification was, most importantly, about selling products: from washing machines to radios and especially refrigerators because of the amount of power they required (Kline 2000; Tobey 1996).

Rural electrification moved slowly in the teens and early twenties, and the rural-urban divide was unmistakable. As a result, the early 1920s saw the start of a debate about a publicly owned national electric system, influenced by the creation of the publicly owned Hydro-Electric Power Commission of Ontario in 1906. The National Electric Light Association (NELA)—the lobby arm of the US electrical industry—immediately attacked the idea of public power. It even went so far as to create its own rural power advisory group and a separate research arm, the Committee on the Relation of Electricity to Agriculture (CREA) in 1923 (Kline 2000).[4]

The battle thus began between those advocating for public power and those advocating for private power. "For progressive reformers," Tobey (1996) writes, "there was no mystery in the failure of private utilities to bring electrical modernization to the majority of American homes in the 1920s. The utilities charged too much for electricity" (40). Progressive enthusiasm reached new levels during a Federal Trade Commission investigation into the so-called power trust in the 1920s. While the investigation found no definitive evidence of collusion, "it compiled eighty volumes of damaging evidence about the NELA's underhanded propaganda methods and serious financial misdealings by the pyramid of holding companies that had come to dominate the electric utility industry" (Kline 2000, 133).

At the state level, Governor Gifford Pinchot of Pennsylvania had taken note of the inequalities in availability (connection) and accessibility (cost) and charged Morris Cooke to investigate the possibility of universal electricity in Pennsylvania. This became known as the Giant Power plan, which began in earnest in 1923. Both Pinchot and Cooke favored public-oriented power plans, and both were passionate about rural electrification. Both also wanted to extend this ideology to the federal level (Brown 1980). Pennsylvania was already well connected, but as Cooke (1948) recalls in his reflections on the plan, many farms used diesel, and those that used grid power paid exorbitant sums and had to assume the entire cost of connecting the farm. In 1925, the Giant Power Board submitted its report and recommended a dual system of public regulation of private power and the construction of publicly owned power plants that would generate power through recycled coal waste (Brown 1980). The plan was defeated in 1926, thanks in large part to private industry opposition, most notably from the NELA (which, incidentally, had its own, fantastically named SuperPower plan) and its research arm, the CREA, which "funded elaborate studies, aimed at high-income farms, that tended to keep agricultural engineers and home economists in the orbit of the utility companies" (Kline 2000, 136). In fact,

> at every turn, private utilities thwarted progressive reformers' efforts to obtain social benefits through regulation. In addition, the conservative side of the progressive ideology, stressing individualism and competition, made it difficult to stretch the justification of reform enough to include the social welfare of the household. (Tobey 1996, 45)

Big Power was doing everything it could to ensure its oligopoly on American power.

Another key development in the 1920s debate over public power was the Muscle Shoals debacle. Here, public power found another champion in Senator George Norris of Nebraska, who lobbied vehemently against Henry

Ford's plan to buy the Muscle Shoals Dam in 1921. Norris wanted the federal government to assume control over the dam and get into the power generation and distribution business. Norris, like Cooke and Roosevelt, saw universal electricity, and rural electrification specifically, as the shibboleth for social modernization of the country (Tobey 1996). States had also entered into the fray, and by 1931 twenty-five had conducted 211 investigations into the possibilities of public power. The climate for public power had changed dramatically in the previous three decades.

While one certainly does not want to fall victim to the trope of the great white men of history, President Roosevelt's role in the electrification of rural America cannot be overstated. As New York's governor, he had pondered the value of public distribution of electricity, endorsed municipal electricity, fought the growing electric power cartels by empowering New York's electric commission with sufficient regulatory powers, and spoke on the need to electrify rural communities. As he remarked in his 1932 presidential campaign, "[Electricity] is no longer a luxury. It is a definite necessity. . . . It can relieve the drudgery of the housewife and lift the great burden off the shoulders of the hardworking farmer" (qtd. in Brown 1980, 34). When Roosevelt became president in 1933, the Great Depression had ravaged the country for almost half a decade, and it was time for change. This change was the New Deal. For rural electricity, it came in the form of the TVA and REA.

The TVA was created in 1933, taking over the Muscle Shoals Dam and generating power that it distributed to local cooperatives, which, incidentally, it also helped form. The TVA proved how wrong utilities and manufacturers were in their claim that electricity was "inelastic." As a reminder, the claim of an inelastic market was used to justify utilities' and manufacturers' market strategies, bottom lines, and anti-regulation stance. In contrast, TVA proved four points to industry and to the country:

> That even households with extremely low ability to pay could modernize their homes and their lives.
> Private utilities could profit from mass social modernization.
> Social modernization could come for the mass of households, who had low incomes, before they had accumulated the savings needed to buy appliances and modernize their homes by cash purchase.
> Disprove the assumption underlying the private electric industry's strategy of domestic electrification in the 1920s, according to which the utilities and manufacturers segmented the home electrical market by income. The

electrical industry targeted the highest income households, constituting about one-fifth of all families, to receive the benefits of appliances. (Tobey 1996, 119–120)[5]

Key, of course, was keeping prices down—something power companies refused to do, or at least refused to lower them to comparable rates of urban users. The TVA demonstrated that demand for electricity and for electrical appliances in rural America was indeed elastic, far more dependent on rates than industry attested. Tobey (1996) links rates to the parallel push toward home ownership and personal equity, arguing that "electrical modernization was not about consumption of goods and services; it was about capitalizing the household with assets that more than paid for themselves. The utilities had failed to understand this point" (121). With the TVA's success secured, and "little TVAs" in the form of power projects and flood-control systems popping up across the country, the stage was set for a full-scale federal push for universal electrification in the form of the REA.

When Roosevelt was elected president, Cooke joined him in Washington as the director of the Mississippi Valley Committee. He later received permission to assess the feasibility of a federal rural electrification program, which manifested in a report titled *This Report Can Be Read in 12 Minutes*. In it, Cooke outlined his vision for a connected America. He argued forcefully for both the need for rural electrification and the need for federal intervention, stating unequivocally that "unless the Federal government, assisted in particular instances by State and local agencies, assumes an active leadership and complete control only a negligible part of this task can be accomplished" (qtd. in Pence 1985, 62). Rural electrification would bring modernization not only to American farm production but also to the people who worked the farms: "Both for the farmer and his wife the introduction of electricity goes a long way toward the elimination of drudgery" (qtd. in Pence 1985, 62). Cooke requested of Congress "an allotment of $100,000,000 [approximately $1.8 billion in 2019 dollars] actually to build independent self-liquidating rural projects" with the goal of electrifying over 5 million farms. At the time, only one in nine American farms had electricity, compared to 95 percent of farms in Holland, 90 percent in France, and 85 percent in Denmark, all of which benefited from sizable government investment (Nye 1990, 314).

On May 11, 1935, Cooke's vision became reality when Roosevelt signed Executive Order 7037 and created the REA with an allocation of $100 million and a mandate to connect the unconnected in rural America. Cooke became its first administrator. Crucial to this experiment was the construction of new lines and the standardization of rates. As Tobey (1996) recounts,

> The progressive political tradition envisioned electrical modernization as a major element of the social modernization of the American home. The president wanted the New Deal to break the cycle whereby high private utility rates and high appliance prices prevented mass electrical modernization. (112)

The REA was set up as a loan agency that would provide self-liquidating loans to power companies and households. In other words, the loans would pay for themselves through the revenue generated by the power companies. Loans expired in twenty-five years.

In its first year, the REA struggled, bound as it was to requirements of being a temporary "relief agency" and because it had no base of operations (Brown 1980). One of the early challenges for the nascent agency was the issue of to whom to loan money. Cooke envisioned a democratic process, fully prepared to fund "private utilities, state power districts or municipal plans, farm cooperatives, and the federal government" (Kline 2000, 142). The utilities had different plans and took aim at the entire $100 million. Under pressure from both Senator Norris and President Roosevelt, Cooke stated that the REA would "give preference 'to applications from public bodies and farm cooperatives,'" thereby neutering Big Power's plans (qtd. in Kline 2000, 144). As Brown (1980) explains, "Cooperatives had a history and tradition . . . that were advantageous. American farmers had long experience with them in the area of marketing, and those electric co-ops in operation as far back as 1913 gave credence to their use" (53). Eventually, Cooke approved loans to eleven cooperatives at an interest rate of 3 percent and a twenty-year amortization.

On May 20, 1936, the REA became a permanent, rather than an emergency, agency through the passage of the Rural Electrification Act. It received an initial appropriation of $50 million for the first two years, which would act as principal for the loans. It would then receive $40 million per year for the next ten years (Brown 1980). Loans would be made at a rate of 3 percent, with payment due in full after twenty-five years. "Loans could be made to finance construction and operation of generating plants, transmission lines, and distribution lines for those in rural areas without central station service" or "for home wiring and purchase of electric appliances and plumbing" (REA 1966, 5). The REA's first loan as a permanent agency occurred on

October 1, 1936, for $25,000 to Meeker Cooperative Light and Power in Litch-field, Minnesota (REA 1966).

Two challenges plagued the early years of federal investment in rural elec-trification. First, farmers and rural residents were not organizing themselves into co-ops at the rate expected by the REA. This prompted the agency to create a "Development Division" and begin rural electrification field trips to promote cooperatives and electrification. The traveling field agents that staffed these trips eventually became known as "agents of modernity" (Kline 2000) and consisted of engineers and home economists. Their rural elec-trification tours, or "Farm Equipment Tours," were dubbed the "REA Cir-cus." From 1939 to 1941, these agents of modernity visited twenty-six states, where "farmers and their families flocked to the circus by the thousands eager to learn more about electricity and the many labor-saving tasks it could perform" (Pence 1984, 118).

The second challenge was an outright revolt by Big Power. Once the cooperatives had been finally organized and funded, "many power compa-nies waged war on the co-ops by building 'spite lines', in the terminology of the REA, across co-op territories to cut them up into unprofitable pieces" (Kline 2000, 146). Said differently, large power companies undercut the nascent cooperatives by wiring parts of the co-op's territory and offering these potential customers discounts. Cooke and his deputy, John Carmody (who would become administrator in 1937), enlisted state organizations to establish, promote, and organize cooperatives and combat these spite lines.

At this time, electrical cooperatives did not generate their own power but rather purchased power from central stations and distributed it to mem-bers. That changed in 1938, when a "Federal court upheld the right of five distribution co-ops to organize a federated cooperative to generate electric power for members" thus cutting out Big Power entirely (REA 1966, 53). Eventually, Big Power backed off its dogged pursuit of cooperatives and began instead to figure out how to profit from the massive federal invest-ment in power.

In 1938, after just two years into formal existence, the REA had funded 350 cooperatives in forty-five states, which in turn were providing electric-ity to 1.5 million farms (Malone 2001). In 1939, that number jumped to 500 co-ops and 180,000 miles of new lines (Kline 2000). That same year, the REA ceased operating as an independent agency and was transferred to the USDA, under the auspices of the secretary of agriculture.[6] As the REA

itself noted, this did nothing to slow the pace of connectivity: "The electrification of rural America continued at an even faster pace. By 1940, loans
totaled $268 million," and its millionth customer had been served. In 1942,
the agency loaned out $100 million in a single year (REA 1966, 18).

The onset of World War II, and America's subsequent involvement therein,
slowed the progress of rural electrification (including the cancelation of the
REA Circus in 1941), but two important events occurred during the war years.
First, in 1942, the NRECA was formed to represent the interests of the hundreds of electric cooperatives created by the REA. Second, Congress passed
the 1944 Agriculture Organic Act, which "established a fixed interest rate of 2
percent for REA loans and extended the repayment period to 35 years" (Buckley 1999, 18). Following the precedent set by the 1934 Communications Act,
this act also embedded the concept of universal service into the mandate
of the REA through the principle of "area coverage," that is, "the practice
of extending service to everyone in a given area who wants it regardless of
how far they live from the main facilities [and] considering the feasibility of
a project as a whole, not based on the cost of individual extensions" (Buckley
1999, 18).[7]

As historian Ronald Kline (2000) recounts in his masterful book *Consumers in the Country*, area coverage, or universal service, had long been REA
policy, but it went unrealized as long as so many "relatively prosperous areas"
needed to be served first. "Once co-ops had accomplished this task," however,
"many were understandably reluctant to extend their lines into less affluent
areas," serving only to replicate the market failures of private enterprise and
once again necessitating government intervention (Kline 2000, 225). In 1950,
the REA made area coverage mandatory for all loan applicants.

The REA was a massive federal intervention into American infrastructure, and it proved remarkably successful: "In less than a decade, from 1945
to 1954, the percentage of farms that were electrified grew from 48 to 93
percent. The figure stood at 96 percent in 1956, the last year data were compiled for what was essentially a completed task" (Kline 2000, 219). This was
accomplished through the creation of 927 cooperatives, which served more
than 2.5 million customers. This history demonstrates the capacity and
capabilities of the federal government to connect the unconnected when it
has a concerted and dedicated plan in place. It also highlights the importance of local champions of connectivity, an element that would become
vital to rural broadband some seventy years later.

Rural Telephony

Based on its success with electricity, telephony was added to the REA's stable of loans on October 28, 1949, when President Harry Truman affixed his signature to the amended Rural Electrification Act, authorizing the REA to "make self-liquidating loans at 2 percent for up to thirty-five years for the extension and improvement of rural telephone service" (Buckley 1999, 21). This act was strongly supported by the NRECA, which had formed a Communications Committee in 1947. Unlike electricity, rural telephony had developed alongside urban and industrial telephony, but its diffusion had stalled and even retreated after the Depression. At the time of signing, rural communities were served by "independent" phone companies—"independent" meaning separate from the monolithic AT&T and Bell networks that crisscrossed the country. Cooperatives were also prevalent, and the history of telephone cooperatives actually predates the history of the REA. Like electricity, however, as we will come to see, the financial security provided by the REA made all the difference in universalizing the telephone.

The story of rural telephony is not one of AT&T—the great octopus of monopoly capitalism—but rather of local communities and farmer cooperatives (Fischer 1987a, 1987b; Kline 2000; MacDougall 2014).[8] We are taught that Alexander Graham Bell invented the telephone in 1876 and that diffusion of the telephone was largely a local affair. That is to say, while Bell owned the patents to the telephone until 1894, it depended on franchises around the country to make the literal connections (Fischer 1992). Still, the Bell Company originally planned for the telephone to be sold and used primarily by businesses and wealthy urbanites, with scant attention paid to residential and rural connections (Fischer 1992). This did not last long, however. Following an initial swell of usage, demand exploded after the expiry of the patents and gave rise to the possibility of non-Bell-affiliated telephone networks. As Steven Keillor (2000) ironically remarks, "Like 1990s decisions about where to lay fiber-optic cables to provide access to an information superhighway, 1890s decisions over access to phone systems were key battles for power between entrepreneurs and cooperators and between rural and urban areas" (237). After Bell's patents expired, independent companies began to provide service to communities either neglected by Bell or in competition with Bell. In 1902, only eight years after patent expiration, 6,000 farmer lines and cooperative mutual systems were up and running alongside commercial company lines

(Fischer 1987a). The number of farmer lines jumped 300-fold in five years, totaling 18,000 in 1907. Rural residents were becoming connected almost as quickly as their urban counterparts.

This diffusion of telephony bucks the trend that sees growth in urban centers before spreading to rural communities. Indeed, unlike the spread of electricity,

> we see a startling reversal of the commonplace that modern technologies spread from the advanced sectors to the backward ones. Whereas most "modern" innovations are last adopted in rural places, for a time, more American farm families had telephones than did town families. (Fischer 1987a, 5)

This was specifically the case in 1920, when 39 percent of farm households had a telephone, in comparison to 33 percent of urban family households (Fischer 1987a; see also Kline 2000, 27, on this point). In the Plains region, an astonishing 80 percent of farms had a telephone!

While the history of telephony inverts the modern narrative of the diffusion of technology (rural regions were connected at the same pace as urban regions), the actors are strikingly similar, and the rural cooperative was essential to the diffusion of telephony in rural America. Telephony, Keillor (2000) argues, lends itself well to the cooperative ethos. Unlike the individual-use technologies of the automobile or radio, the telephone is based on a network effect: the more people who had a telephone, the more users benefited from the connections. This is perfect for a cooperative.

There were actually three types of non-Bell telephone systems:

1. Major commercial companies (the "independents")—often serving the larger rural towns unserved by the Bell network
2. Small mutual companies, or "stock mutuals," set up by users to serve themselves wherein members paid a subscription fee and had access to a switchboard
3. Farmer lines—simple cooperative systems, often built without switchboards, where costs were shared, local materials used, and users performed their own repairs but paid no subscription fee (Buckley 1999; Fischer 1987b)

Independent companies typically served small rural towns neglected by Bell but reserved their service for the profitable and denser township, rather than the remote farmer (MacDougall 2014). While some independents served farms, the majority of farmers took it upon themselves to connect their communities. Farmer lines were often tackled as DIY projects, many times using materials already at hand, including barbed wire or chicken wire, leading to unsteady connections, but connections nonetheless. Farmer lines had upward of twenty users connected to the same line, which earned them the name "party lines" (Fischer 1992) and brought with them the issue of "listening in" and moral panics over gossip (Kline 2000). The Midwest, with its long history rooted in populism and communitarianism, housed the majority of these nascent telephone cooperatives (Keillor 2000; Mac-Dougall 2014).

In some instances, non-Bell companies competed in the same market as Bell. Their lack of interconnection forced some users to purchase two lines and two phones—one to call those on the non-Bell network and one to call those on the Bell and AT&T (long-distance) networks (MacDougall 2014). This became known as "dual service."[9] For its part, Bell was incredibly aggressive, pricing out competitors, refusing to sell equipment, refusing to connect with non-Bell networks, or simply buying out competitors (Fischer 1992; MacDougall 2014). Despite Bell's best efforts, however, telephone cooperatives quickly became a mainstay in rural America. By 1912, as Linda Buckley (1999) reports, "there were more than 32,000 rural telephone systems, and by 1917, the number had reached 53,000" (14). The number peaked in 1927 with 60,000 rural telephone systems. These rural networks offered cheaper rates (or none at all) and low initial financial investment for cooperative members. As MacDougall (2014) quipped, "One Michigan independent described his no-frills rural telephone as 'cheaper than the Bell' but 'better than walking'" (144).

Rural telephone cooperatives and stock mutual companies, with their local focus and DIY ethos, succeeded where Bell stalled or failed: to connect those whom the "free" market left unconnected. "Like the early cooperatives, which drew on the rhetoric of populism to fight the perceived injustices of middlemen unfairly marketing farm products," Kline (2000) concludes, "the telephone co-ops struck a blow against another monopoly, the Bell 'octopus'. Independents responded by naming many of their companies the 'Home' or 'People's' Telephony Company" (29). As MacDougall (2014) demonstrates in his fascinating history of the telephone in North America, these "people's networks" posed a very serious threat to the Bell monopoly—that is, until the Depression.

Between 1920 and 1940, approximately 1 million farms abandoned their telephones, representing a one-third reduction in rural telephony (Fischer

1987b). Obviously, the depression of household income was a determining factor in this decline (Fischer 1987b; Kline 2000). Kline (2000) suggests that another reason why farmers were so quick to abandon their phones during and after the Depression was because it ranked low on the hierarchy of needs for material goods. "When deciding what expenses to cut during the hard times of the depression," Kline (2000) writes, "many farmers chose to give up their phone but keep the car" (227). In other words, the telephone was more of a luxury for rural Americans "compared with the multipurpose automobile" or the radio (228).

Another reason for the rural abandonment of telephony was that the rural telephone line suffered from interference from, ironically enough, rural electric wires, wherein a "hum could be heard on the telephones hooked up to lines near (or often on the same poles with) power lines" (Kline 2000, 228). On top of this, the lines were degrading, having been assembled on an ad hoc basis and unable to withstand the years and advances in technology (Fischer 1987b). Fischer (1987b) notes that in post–World War II Indiana, "mutual companies' lines were out of order on an average of 13 days a year, and in some cases for weeks" (301).

A final explanation for the lack of rural telephony during and after the Depression is simply industry neglect. Rural customers were best served during the years of competition, after the patent expiration in the mid-1890s but before the 1913 Kingsbury Commitment. The Kingsbury Commitment was an agreement between the US Department of Justice and AT&T that divorced AT&T from Western Union, slowed the progress of AT&T's mergers and acquisitions of other telephone systems and forced it to open its network to independent competitors. The independents that connected to Bell's network, especially in rural areas, became more popular than their local competitors in dual-service areas and thus drove them out of business. Thus, the Kingsbury Commitment effectively cemented local telephone monopolies and ended dual lines and local competition (MacDougall 2014).[10] These remaining companies—the reigning Bell companies and major independents—went on to ignore rural areas and remote farmers, preferring instead to focus on the lucrative urban markets (Fischer 1987b). These issues, compounded by the financial constraints imbued by the Depression, no doubt contributed to rural telephony's lack of resiliency even after the Depression. As Fischer (1987b) reports, "[Rural] telephone subscription dropped from 39% in 1920 to 34% in 1930 and to 25% of all farms in 1940" (297).

Despite the decline in subscribers, many telephone cooperatives sur-
vived the Depression, even with as few as fifty (or fewer) members. In 1937,
the US Census Bureau reported 33,347 telephone cooperatives in the country, with average membership hovering around twelve subscribers (Kline
2000). While the lines may have existed, many had not been updated or
well-maintained over the years, leading to another form of communica-
tory divide, not unlike the difference between satellite broadband in rural
areas and fiber connections in towns and cities: all have internet, not all
have broadband. The same can be said for many of these rural lines—all
had "telephone," not all had connectivity. To correct this deficit and further encourage telephone adoption among rural communities, Congress
proposed the creation of a Rural Telephone Administration, akin to the
REA, to provide low-interest loans to spur rural connections. The first iteration appeared in 1944–1945, sponsored by Alabama senator Lister Hill, who
noted at the time that only 5 percent of Alabama farms had a telephone.
The bill never made it past the House, and it would be another half decade
before it came to fruition.

AT&T opposed the initiative, stating its own rural telephone program
would suffice for rural America. The giant octopus even went so far as to
partner with the REA to conduct telephone-over-power-line experiments
in rural areas and worked with electric co-ops to permit Bell wires to be
strung on co-op-owned poles. Echoing Bell, the US Independent Telephone
Association argued that any rural telephone bill would "instantly threaten
and ultimately completely destroy these six thousand small independent
businesses in 11,000 cities, villages, and communities, providing service in
almost 70 percent of rural areas in the United States" (qtd. in Hadwiger and
Cochrane 1984, 229).

One of the key points of disagreement was whether farmers and rural
residents even wanted the telephone (a debate not unfamiliar to those in
electricity and broadband). "Industry critics and farm representatives," writes
Claude Fischer (1987a), "argued that rural Americans desperately wanted telephone service (at reasonable rates)" (13). Demurring, the major telephone
companies "contended that farmers did not really want service . . . did not
really need the telephone" and could not afford the markup necessary to provide service to rural areas (Fischer 1987a, 13). In short, the telephone industry, like the power industry before it, saw rural service as inelastic. Congress
was unconvinced by this argument. In a line that could have come from a

speech by Roosevelt on electricity or a contemporary hearing on broadband, the 1949 House Committee on Agriculture's report on *Rural Telephones* stated bluntly: "We cannot expect farmers to be satisfied with living conditions that are 20, 30 or 50 years behind those in the city, as they are in the case of telephones" (qtd. in Hadwiger and Cochran 1984, 222).

With the support of the major farm groups, the NRECA, and the REA, Congress passed the Telephone Act (also called the Hill-Potage Act) in 1949. This amended the Rural Electrification Act to permit the REA to provide low-interest loans for rural telephony. Cooperatives were eligible for these loans, but public utility districts were not. Unlike the loans for electricity, "the agency required substantial equity for telephone loans, often twenty-five to fifty dollars from subscribers, rather than the nominal membership fee of five to ten dollars for those signing up for electricity" (Kline 2000, 233). In addition, the REA was instructed to give preference to previously established telephone cooperatives rather than new entrants. Despite these barriers, the program flourished. As Hadwiger and Cochran (1984) recount, "Within three months after the Rural Telephone Act was passed, the REA had received 1,117 loan inquiries from forty-four states. By June 1950, the REA had made seventeen allocations and 500 applications were on file" (231).

The REA's rural telephone program was an immediate success, with farm connectivity jumping from 25 percent in 1940 to 65 percent in 1959. Interestingly, the first company to win a telephone loan from the REA was not a cooperative. The Florala Telephone Company, in Florala, Alabama, secured a loan of $243,000 to rebuild its outdated system and add 130 miles of new line. Two months later, in April 1950, the Emery County Farmers' Union Telephone Association out of Orangeville, Utah, was the first cooperative to win a loan. A loan to the Iowa Falls Telephone Cooperative quickly followed suit, and it is notable because ten mutual companies combined efforts to win it—a practice that would be repeated often (Bass 2010; Buckley 1999). The honor of the first system going live using REA money goes to the Fredericksburg and Wilderness Telephone Company in Chancellor, Virginia, which used its loan to upgrade its rotary network to a dial telephone system. Hadwiger and Cochrane (1984) applaud the REA for its distinctly "rural" mindset and agenda, "savor[ing] the challenge of reaching remote users" (232). To achieve that, the REA depended on existing telephone and electric cooperatives to proselytize rural telephony. Aiding this was the creation of a new trade association, the National Telephone

Cooperative Association (now the NTCA—The Rural Broadband Association), an outgrowth of the NRECA that achieved independence in January 1954 (Bass 2010).

In 1964, 75 percent of farms had a telephone, but demand was constant, and the REA had to search for new sources of capital. It reached an agreement with President Lyndon Johnson's ever-budget-conscious administration to approve loan rates above 2 percent, in exchange for access to greater financial resources. Upholding its part of the bargain, Congress created the Rural Telephone Bank in 1971, which "secure[d] its funding from debentures purchased by the Secretary of the Treasury rather than from congressional appropriation" and would finance loans based on the "cost of money rate" (Hadwiger and Cochrane 1984, 236). "By 1978," Hadwiger and Cochrane (1984) write, "telephone borrowers . . . were using $425.8 million in long-term financing, of which $159 million was from the rural telephone bank at market rates, another $191.6 million was at the REA's 5 percent rate, and $74.7 million was at 2 percent" (237).

With its goals largely accomplished, the REA spent the next few decades filling the telecommunications gaps. This included focusing on Native American communities and rural communities in Alaska. It also meant flirting with emerging advanced technologies like cable television and satellite telephony (Buckley 1999). In fact, in 1978 the REA entered into an agreement with Commonwealth Telephone Company out of Dallas, Pennsylvania, to fund the "nation's first fiber-optic system for commercial use" (Buckley 1999, 65).

The REA played a decisive role in wiring rural America, with both electricity and telephony. Rural communities and farmers, however, played an equal, if not more important, role than the federal government. Through the development of electric and telephone cooperatives, rural Americans connected themselves to the wires of modernity far more successfully than the agents of mainstream capitalism. Indeed, they did so in spite of Big Power and the Bell system. Robust, dedicated, and purposeful federal policy, in concert with equally dedicated local communities, managed to connect this country at a time when over 40 percent of the population lived in rural areas (US Census Bureau, n.d.). It is this esprit de vie that we are missing today when it comes to rural broadband, but one that this chapter aims to recover.

Conclusion: Lessons Learned . . . and Not Learned

There are many similarities between the history of rural connectivity and
the present conditions of rural broadband. These include the neglect, and
even hostility, of the major incumbent providers to rural America and the
argument by detractors that rural America does not need, or want, or use,
these technologies of modernity. What these parallels teach us is that communication policy and, to be bolder, technological diffusion, is a cyclical
process: the concerns we had in an earlier generation revisit themselves upon
us in the guise of new technologies (Marvin 1988; Mosco 2005). In the case
of connecting rural America, the issues of the rural penalty, market failure,
and the call for and debate over universal service, define the conversation.
The only difference is that today we speak of broadband and yesterday we
spoke of electricity and telephony. By understanding the cyclical, and some
might even argue circular, nature of communication policy and debates
of communication technologies, we can extract six important lessons that
today's rural broadband advocates can learn from their predecessors:

1. Rural connectivity is a market failure engendered by incumbent neglect.
   In the 1930s–1950s, incumbent electric companies and telephone companies ignored rural communities and justified their decisions by claiming a lack of return on investment.

2. An ethos of localism, through community champions, local decision-
   making, and local ownership, is vital to correct the market ontology of
   large providers. Rural communities have a long history of connecting
   themselves when ignored by the larger providers.

3. Cooperatives are key actors in the crusade for universal connectivity. The
   REA was founded to foster electric cooperatives and empower telephone
   cooperatives.

4. No prospective provider should be dismissed from offering service and
   accessing resources; this includes municipalities. While contemporary
   policy eschews competition in favor of incumbency, history records the
   importance of upstarts and cooperatives.

5. It is not enough to talk about *availability*; we also need to talk about *accessibility*, which means talking about rates, prices, and cost, along with literacy and skill development. One of the key lessons of the history of rural

electricity and telephony is that of rates and prices. A second important lesson is the success of the REA Circus to promote adoption and teach literacy.

6. Coordinated, dedicated, and robust federal policy and financial support is crucial. The REA Act and the amendments that followed empowered the REA to craft and implement a successful universal service policy.

The lessons of market failure and federal policy in the list above were not lost on my interviewees in 2017 and 2018. Many pointed to these dual conditions as key determinants in rural electrification and telephony and contrasted these dynamics with the contemporary conditions of rural broadband that see the existence of market failure but the *absence* of federal policy. To the question posed about the reasons why the rural-urban digital divide persists, respondents unilaterally pointed to conditions of market failure, even if they did not use the term itself. Market failures, noted Don Paxton, manager of MINET in Independence, Oregon, are "the most significant, if not the only barrier to be honest . . . just the lack of density to support a sustainable business model of a robust service in the areas that remain to be served" (personal communication, 6/23/17). Jonathan Adelstein, former FCC commissioner and RUS Administrator and now CEO of the Wireless Infrastructure Association (WIA), echoed this statement:

> Because broadband offers so many externalities that benefit communities beyond the carriers that deliver the service, the full value to society is often not captured in rural areas. It is a classic market failure where there isn't sufficient revenue to cover the cost of the network deployment. (personal communication, 10/23/17)

The challenge of rural broadband deployment, continued Adelstein, is economic:

> Rural areas have some similar fixed costs, but smaller populations that are less dense to cover those fixed costs. The lack of density creates higher costs per customer for actually deploying and operating broadband networks. . . . The economics are challenging for private companies that invest in rural networks looking to maximize their return on investment. (personal communication, 10/23/17)

As Adelstein states, the return on investment, or lack thereof, forces for-profit, shareholder-based companies to ponder, "Is it worth it, or profitable, to cover rural areas? Even if it is profitable, it may provide a lower return on limited capital than spending to improve urban networks to avoid losing customers in more dense areas" (personal communication, 10/23/17). Cooperatives need not ask the same questions (see chapter 4).

Adelstein points to the crux of the definition of market failure—the private market's inability to provide "socially necessary goods and services" (Pickard 2019, 65). Broadband, much like electricity and telephony, is not simply a commodity but a socially necessary good—a utility—that the private market is unable to support in rural America. As Bernadine Joselyn, director of public policy and engagement for the Minnesota-based Blandin Foundation told me, "For me, kind of a bottom-line conclusion to this is that the market economy isn't very well suited to the public needs and putting for-profit . . . on top of ensuring basic access has failed." She credited co-ops with changing this mentality: "And that's why co-ops are so appealing, because they have this model of member needs and the common good" (personal communication, 1/15/19).

Respondents like Joselyn point to the need to look beyond the market to solve the digital divide. Progressives tend to look to the federal government to intervene on the public's behalf in the wake of market failure (Ali 2016; Pickard 2014a, 2014b, 2015, 2019; Stiglitz 2009). This was the call heeded by Roosevelt's administration and the REA. Connecting rural America, to put it bluntly, is far too important to the well-being of the country to leave it in the hands of the free market and neoliberal capitalism.

To economists subscribing to the theory of market failure, state intervention is not a novel idea (Bator 1959; Medema 2009). Noted English economist Arthur Pigou made this point in 1935, as summarized by economist Steven Medema in *The Hesitant Hand* (2009):

> What we have here are, in modern language, allocation failure, distribution failure, and stability failure, respectively—"failures" in the sense that, in each instance, the market fails to generate the best possible result for society. When such market failures arise—that is, when "private self-interest, acting freely subject only to the ordinary forms of law, does not lead to the best results from a general social point of view"—there is, says Pigou, "a prima facie case for State action." (68)

In 1935, the failure to electrify rural America was met with a federal government initiative that created the REA to encourage the development and funding of locally based cooperatives. The same thing happened in the 1950s for rural telephony. Many of my respondents argue, and I agree, that a similar concerted, omnibus federal initiative is required for rural broadband. Take Tim Marema, vice president of the Center for Rural Strategies and editor and regular contributor to the *Daily Yonder,* a news website dedicated to rural America. In an interview early in my research, he argued, "There needs to be some thoughtful and strategic and careful market interventions to create a more competitive atmosphere, to make it more attractive to investment and to hold private entities accountable for what they said they would do" (personal communication, 1/18/18). RJ Karney, director of congressional relations for the AFBF, went further and connected the dots between market failure, the rural penalty, state intervention, and electricity:

> What needs to occur is, in order for . . . private companies to build out in rural America, they don't have the economic business model. . . . So there does need to be—just as we did with electricity, just as we did with telephone service—this universal service, that this is a public good, that it is correct to use taxes, user fees, other policy initiatives that can incentivize companies to build out in more remote areas that they might not get the true economic return that they will get in a more populous area, right? And that's one of the biggest barriers is trying to get the infrastructure built out because that is costly. (personal communication, 6/8/17)

Karney is quite right to suggest that broadband, like electricity and telephony before it, is a *public good in the sense that it is good for the public.* Broadband facilitates a multitude of positive transactions, both social and economic. One key benefit, or positive externality, is the network effect or "network externalities": the network gets better as more people use it (van Schewick 2007). I listed other positive externalities of broadband to rural communities in the introduction to this book under the header of "the five pillars of rural broadband." These include precision agriculture, education, health care, job creation, housing values and civic engagement.

As an economic principle, however, broadband is not strictly a public good but rather an "impure" public good or "club good" (Frishmann and van Schewick 2007; Hofmokl 2010; Johnston 2016; Raymond and Smith 2014; Yoo 2006).[11] In economic theory, the term "public goods" describes those goods that are both nonrivalrous (one person's consumption does not diminish the next person's consumption of the same product) and nonexcludable (you cannot exclude those who don't pay for it) (Ali 2016; Baker 2002). News and information are media examples of public goods (Pickard 2019). As it stands in the United States and elsewhere, however, broadband does not meet the criteria of a public good. This is because broadband suffers from network congestion, making it partially rivalrous (Frishmann and van Schewick 2007; Hofmokl 2010; Raymond and Smith 2014; Yoo 2006). Congestion is defined as a "situation in which one individual's consumption reduces the quality of service available to others" (Cornes and Sandler

1996, 9). Congestion is particularly notable on cable networks, where the more users on a neighborhood network, the slower the network will be. In other words, there are moments in which broadband access and network performance becomes rivalrous:

> The physical infrastructure of the Internet [of which *Farm Fresh* describes]—the interconnected networks—is a partially nonrival resource, meaning that it is sharable but congestible. Infrastructures are sharable in the sense that the resources can be accessed and used by multiple users at the same time. Infrastructure resources vary in their capacity to accommodate multiple users, however, and this variance in capacity differentiates nonrival (infinite capacity) resources from partially non-rival (finite by renewable capacity) resources. Infrastructure resources of finite by renewable capacity are congestible. (Frischmann and van Schewick 2007, 393)

Johnston (2016) adds that, in addition to network congestion, broadband has excludable characteristics like subscription cost, which we saw in the introduction is a major determinate of broadband access, particularly in urban areas (Chao and Park 2020). Perhaps with a universal fiber system, congestion will be eliminated, and perhaps with a public system, cost concerns will be eliminated—but neither has happened yet. The prevalence of multiple suboptimum networks—such as cable, DSL, and satellite—renders broadband partially rivalrous; its subscription cost renders it partially excludable.[12]

Public goods and club goods share the problem of market failure (Cornes and Sandler 1996). Indeed, the nonrivalrous and nonexcludable attributes of public goods, and the partially rivalrous and excludable qualities of club goods mean they are difficult, if not impossible, to be supported solely by the private market because people can access them without paying. This is known as the "free rider" problem (Pickard 2019). Just because something is a club good and not a pure public good, therefore, does not obfuscate the government's responsibility to correct its market failure (Cornes and Sandler 1996). This is precisely the case with broadband, and it was the same for electricity and telephony. Broadband's designation as club or public good is less important than its existence as a market failure and the output of positive externalities (Cornes and Sandler 1996). Indeed, Mark Raymond and Gordon Smith (2014) argue that it is precisely because broadband is a club good that governments have an obligation to ensure universal access:

> Maintaining the global reach and interoperability of the Internet, and thus maxi-mizing its value to humanity, requires ensuring that access to these clubs remains open to all, and that restrictions on member behaviour do not exceed the mini-mum requirements of public safety. (22)

This chapter has recalled the role of government intervention regarding two earlier club goods, electricity and telephony. In the 1930s the REA solved the market failure of rural electricity and did the same for rural telephony in the 1950s not only by injecting capital into the market, but also by championing the cause, educating users, fighting off industry challenges, and keeping service local. This was not only about loan programs but also about the REA launching a rural electricity road trip throughout the country, and the creation and sustainability of locally based cooperatives. The analogy with rural broadband should not be lost or underappreciated. Shawn Irvine, director of economic development for Independence, Oregon, said it best: "I mean, because broadband is the next electricity, you know?" (personal communication, 6/16/17).

There is one crucial difference between the history of rural connectivity and its present condition: the role of major corporations. We learned in this chapter that major power companies, and later Bell/AT&T, actively ignored rural people and then vehemently opposed government intervention. Inevitably, however, they acquiesced to the existence of the REA and to the service provided by the nascent co-ops. This is only half true today. The major incumbent broadband providers in rural America, which are more often than not the giant telecommunications companies spawned by AT&T, champion federal government intervention—but only to the point where they can profit. This is the topic of chapter 2. There, we will learn that the major telecommunications companies receive the bulk of federal subsidies and that they have successfully lobbied to secure low build-out requirements. In the 1930s, 1940s, and 1950s, the major providers backed off. Today, they are not giving up without a fight.

The next chapter explains the US government's current policy approach to rural broadband, led by the FCC. It discusses how the current subsidy system, known as the Universal Service Fund (USF), fails to meet the needs of rural Americans, but succeeds at padding the bottom line of incumbents: the inverse of the REA's mission. The following chapter (chapter 3) examines the REA's successor, the Rural Utilities Service (RUS), and its well-intentioned but incomplete role in the modern effort to connect rural America.

2 When Good Is the Enemy of Great: The Four Failures of Rural Broadband Policy

What happens when public policy fails?[1] Policies are established to achieve some socially desirable goal: literacy, protection, a healthy body, a healthy market. Harold Lasswell (1951) defines policy as "the most important choices made in either organized or in private life" (88). To achieve these goals, whatever they may be, we need regulation, which is the "form of government that compels those entities over which it has legal jurisdiction to act or refrain from acting in the manner in which they would otherwise tend to act" (LeDuc 2003, 168). Regulation is meant to uphold and protect the public interest (R. Horwitz 1989). We have regulations mandating seat belts, controlling pollution, assessing corporate mergers, and limiting the wattage of toasters. Regulation constrains the inhumane behavior of private markets and private industry (Baker 2002; Novak 2013). Capitalism, as Karl Marx ([1867] 2004) reminds us, is a predictable beast. Left to its own devices, it will maximize profit and eliminate competition. It has no concern for safety, standards, or labor rights and will inevitably march toward monopoly (Baran and Sweezy 1966).[2] It will also ignore unprofitable scenarios, leading to market failure. Regulation in the public interest is meant to correct for these failures.

Communication policy and regulation are special (Napoli 2001). They are special because the communication industries are special. They are not simply a profit-maximizing industry and contributors to the national economy. (And they are that: Amazon, Apple, Google, and Microsoft were the most valuable companies in the world in 2019 [Handley 2019].) They represent our window to the world. As such, the regulation of this democratically vital set of industries takes on both economic and social imperatives (Napoli 2001). Broadband, as we recall from the introduction to this book, facilitates economic development, education, health, civic engagement, and quality of life—what I called the five pillars of rural broadband. For rural communities, the regulatory imperatives of economic and social well-being are even more important, as broadband corrects the deficits wrought by the rural penalty (Nicholas 2003; E. Parker et al. 1989).

As long as broadband provision remains the exclusive purview of the private market, full rural broadband deployment is impossible without government intervention. As Kyle Nicholas (2003) writes, "In the era of the internet, the ability of rural people to communicate, whether for the purposes of commerce or community, is still shaped by public policy" (287). Chapter 1 demonstrated how, when it comes to rural connectivity, private companies refuse to connect sparsely populated communities because there is no return on investment—the very definition of market failure. To remedy this in telecommunications, Congress established a policy goal of "universal service," first with telephony in the 1934 Communications Act, then electricity in the 1944 Agriculture Organic Act, and later with "advanced telecommunications" in the 1996 Telecommunications Act.[3] Here, Section 254 established the principle of universal service and directed the FCC to create an explicit subsidy to ensure all Americans have access to advanced telecommunications services:

> Consumers in all regions of the Nation, including low-income consumers and those in rural, insular, and high cost areas, should have access to telecommunications and information services, including interexchange services and advanced telecommunications and information services, that are reasonably comparable to those services provided in urban areas and that are available at rates that are reasonably comparable to rates charged for similar services in urban areas. (§254(3))

Section 706 of the act requires the FCC to report annually on the progress of the deployment of advanced telecommunications (i.e., broadband), and more generally, to

> encourage the deployment on a reasonable and timely basis of advanced telecommunications to all Americans (including, in particular, elementary and secondary schools and classrooms) by utilizing, in a manner consistent with the public interest, convenience, and necessity, price cap regulation, regulatory forbearance, and measures that promote competition in the local telecommunications market, or other regulating methods that remove barriers to infrastructure investment. (§706(a))[4]

After the passage of the 1996 act, the FCC began to draft the regulatory architecture necessary to subsidize telephone service in high-cost areas. The Universal Service Fund (USF), administered by the Universal Service

Administrative Company (USAC), was borne of this process. In 2011, following the recommendations of the 2010 NBP, the USF was transformed to subsidize both voice and internet connections, thus beginning the country's aim of universal broadband (see FCC 2010, 2011b).

Today, the subsidization of rural broadband deployment is articulated within two complicated subsidy systems, one administered by the FCC through USAC and the other administered by the RUS under the USDA. This chapter focuses on the broadband subsidy programs administered by the FCC and USAC, which distribute over $8 billion annually. Chapter 3 then focuses on the RUS subsidy programs.

I argue in this chapter that if communication policy exists to mitigate market concentration *and* enhance the public good, then FCC regulations targeting rural broadband deployment have failed to deliver on both accounts. Instead, the regulations that govern rural broadband serve the interests of the market and specifically the interests of the largest telecommunications companies.

FCC rural broadband policies have failed in four capacities: *management*—planning a rural broadband strategy; *meaning*—setting the definitional boundaries of broadband and broadband technologies; *money*—incentivizing telecommunications companies to connect rural communities; and *mapping*—measuring broadband deployment across the country and publishing these findings. These policy failures occur because of what I call the politics of "good enough." This politics permeates FCC policy and is used to justify the use of inadequate speed definitions for broadband, the insufficient reporting requirements of ISPs, the deployment of subpar technologies, and the distribution of millions of dollars a year to the largest telecommunications companies in return for poor connectivity. Through my discussion of the four failures of rural broadband policy, I argue that those pushing the politics of "good enough" are the major legacy and incumbent telecommunications companies and satellite companies, like AT&T, CenturyLink, Windstream, Frontier, and Viasat, which benefit from billions of dollars of subsidy but are not required to upgrade their networks because of low buildout requirements and substandard speed thresholds. As one respondent said to me in an interview, "Nobody ever got in trouble at the FCC for supporting the telephone industry."

So, what happens when public policy fails? The market runs amok, and rural communities are left unconnected.

Policy Failure

At its broadest, policy failure or regulatory failure occurs when established regulations fail to accomplish a stated policy goal (Baldwin et al. 2012; R. Horwitz 1989). For instance, regulation that is meant to uphold the public interest fails when it is captured by industry interests, something Robert Horwitz calls the "perversion of the public interest" (27). Adding to this, critical policy scholars like Victor Pickard and even mainstream economists like Joseph Stiglitz relate policy failure and market failure: "failure to act in the face of market failure . . . amounts to 'policy failure' especially from a public interest perspective" (Pickard 2013, 339). Victor Pickard goes on to suggest that policy failure occurs because of what he calls policy's "market ontology": policymaking in the United States is gripped by a neoliberal ideology that comes with an unshakable belief in the salience of the free market. As Stiglitz (2009) explains of the 2008 global financial crisis, "The primary reason for the government failure was the belief that markets do not fail, that unfettered markets would lead to efficient outcomes and that government intervention would simply gum up the works" (19–20). A paradox therefore exists in the world of contemporary communication policymaking. On the one hand, neoliberal ideology demands a free and unfettered market based on the assumption that it will bend to the will of the rational consumer (think Adam Smith's "invisible hand"). On the other "hand" (pun intended), we have market failures, which illustrate the fallibility of the market and of the market ontology. These demand regulatory intervention. In this way, political economist Robert McChesney (2003) was quite right to note that the struggle over regulation is not one between regulation and the "free" market, but rather between regulation in the public interest and regulation in the market's interest. Rural broadband fits squarely within this paradox.

Pickard's market ontology works in a second capacity as well—neoliberal ideology assumes not only that the private market will cede to rational consumer demands, but also that the private market is the only vehicle capable of providing a service. This belief underscored legislation in the nineteen states that prohibit or inhibit municipalities from starting or expanding their own broadband deployment projects (Baller Stokes 2019; Chamberlain 2019; see chapter 4). The end result is a policy goal and a regulatory system serving two different masters. Said differently, we have a series of policies and regulations aimed at "serv[ing] the interests of monopoly capital" rather than the public interest and the public good (Baran and Sweezy 1966, 65). In the case of broadband, our current policy architecture has failed to mend the rural-urban digital divide but succeeded in bolstering the bottom line of the country's largest telecommunications companies.

The Four Failures of Rural Broadband Policy

Management: The 2009 Rural Broadband Strategy

Section 6112 of the 2008 Food, Conservation, and Energy Act, known colloquially as the 2008 Farm Bill, ordered the FCC and USDA to research and publish a national rural broadband plan, which they did in 2009 (Copps 2009). This was the first and only time that the United States had a dedicated plan for rural broadband deployment. Unfortunately, it was short lived, eclipsed the following year by the 2010 NBP, which took the focus away from the rural and aimed it toward the nation's digital agenda (FCC 2010). The 2009 plan, however, offered a distinct opportunity to craft a vision of rural America's digital future.

The FCC spearheaded the initiative and opened up a docket for public comment on March 10, 2009. While the public was initially given only an anemic fifteen days to comment, the commission received 225 comments from a variety of stakeholders. These ranged from individuals to municipalities and cooperatives to the largest players, including AT&T, USTelecom, Microsoft, and Google. The bulk of the commenters, however, were small ISPs and their trade associations.

Typically, in public comments filed to the FCC, there is disagreement between public/consumer interest commenters, who tend to be pro-regulation (as long as it is regulation to protect consumers), and industry commenters, who tend to be pro–free market and against regulation. In this docket, however, the discord fell between local and national providers. While all agreed that rural broadband is important and most agreed that it is a market failure, they disagreed about how to correct this failure. Key themes included interagency cooperation, mapping, subsidy, technological neutrality, and localism.

The most prominent theme was interagency cooperation. Commentators strongly encouraged greater cooperation and data sharing between the FCC and USDA specifically and between the NTIA and other departments as well. There was disagreement, however, about which agency should spearhead the rural broadband strategy. The AFBF, for instance, preferred USDA while AT&T preferred the FCC (see chapter 3). Mapping and data gathering was another key theme, no doubt precipitated by the Notice of Inquiry that singled out this issue and by the American Recovery and Reinvestment Act of 2009 (also known as the Recovery Act), which charged the NTIA with creating a National Broadband Map and the FCC with developing a national broadband plan (§6001). Subsidy was a third theme. Many agreed that the FCC's USF needed reformation so as to include broadband in addition to, or in replacement of, telephone subsidies (this, of course, did occur as a result of the NBP). Technological neutrality was another prominent theme. Many were concerned that the rural broadband report would encourage specific technologies (e.g., fiber) over others or would impose "gold standard" speeds, which would by default eliminate certain technologies (e.g., satellite).

The final theme was local communication. Many public interest and local and cooperative ISPs made it clear that rural connectivity is a fundamentally local issue, one best served by local providers. The National Association of State Utility Consumer Advocates (2009, 2) made the provocative point that "any rural broadband strategy that considers telecommunications carriers must recognize the distinction between rural areas of the country that are served by smaller rural telephone companies and the rural areas served by the larger, 'non-rural' companies, including AT&T, Verizon and Qwest." At the time, federal subsidy programs favored incumbent (i.e., national) providers. The same is true today. Local providers are eclipsed by their national counterparts. Access Humboldt, both in its individual filing and as a member of the public-interest consortium Rural Broadband Policy Group (2009, 3), echoed the need for a local approach: "Local ownership provides self reliance, and local investment in community. Absentee-ownership of broadband infrastructure has failed to serve rural communities. Local and public ownership of communications infrastructure will support more sustainable community network investments." This local and place-based solution for rural broadband is a key element in solving rural broadband. Indeed, as this group continues,

> The current market-driven policies for the build out of broadband do not adequately serve rural communities. After all, the federal government defines rural areas as regions lying outside metropolitan markets. Therefore, market-driven solutions for rural areas are problematic by definition. (Rural Broadband Policy Group 2009, 3)

Many of the state associations agreed, wanting greater authority in the funding programs for rural broadband.

The FCC's *Bringing Broadband to Rural America: Report on a Rural Broadband Strategy* (Copps 2009) echoed many of the aforementioned themes, though it downplayed the idea of local ownership. The report is written like no other FCC document—in the first person, voicing the comments, suggestions, and recommendations not of the commission but of acting chair Michael J. Copps. Copps was one of the most progressive commissioners ever to sit on the FCC (Pickard and Popiel 2018). His imprint is found throughout this document, beginning with the repeated acknowledgment that rural connectivity is not possible without government intervention. This was true with electricity, telephony, and the highway system as it was with the invention of the internet and now universal broadband. There is also an unabashed acknowledgment of market failure consistent with Copps's Keynesian epistemology:

> Although the free market has many benefits, such as driving down the costs of services for consumers and improving service quality, it also can leave behind geographic areas with high costs and lower profit potential. Such is the case with many rural areas. . . . Therefore, we believe that government action is needed to encourage deployment of broadband for rural areas. (Copps 2009, 50–51)

The report also highlighted other difficulties of rural broadband deployment, including topography, weather, and distance (both in terms of extended last miles and distance of middle-mile connectivity). Because these difficulties require a community-by-community approach, the report took a technologically neutral philosophy, albeit one with scalability and durability in mind:

The solutions for rural broadband should reflect consideration of the full range of

> technological options available, and should not elevate the need for short-term progress over longer-term objectives. Rural broadband likely will include a variety of different technologies that together can support the state-of-the-art, secure, and resilient broadband service that should be our goal for rural America, just as it is for the non-rural parts of the nation. However the rural networks are configured, they should be designed on principles of durability, reliability, openness, scalability, and interoperability so that they can evolve over time to keep pace with the growing array of transformational applications and services that are increasingly available to consumers and businesses in other parts of the country. (Copps 2009, 4)

While no doubt meant to provide the greatest flexibility in rural broadband deployment, the belief in technological neutrality, as we will see, ended up as a crutch to justify subsidizing inadequate technologies, rather than a tool of rural community empowerment and choice.

By every measure the goals and recommendations listed in the report, which included greater coordination, the assessment of rural broadband needs, crafting a plan for deployment, and progress monitoring, have gone unrealized in the decade since its publication. The country still grapples, for instance, over middle-mile access and cost, spectrum scarcity, access to poles, and the placement of towers, all of which were noted as challenges in 2009. The one recommendation that was fulfilled was the change to the USF, although even this is a work in progress. Moreover, the transformation of the USF was most likely done at the behest of the NBP and not the rural broadband strategy.

Bringing Broadband to Rural America was released without fanfare and was ultimately eclipsed by the 2010 NBP, in terms of both scope and visibility. This was even acknowledged in the preamble text: "In some respects, events overtook this effort as Congress provided new direction and support for federal broadband policies and initiatives, guidance which frankly has reshaped our approach to the development of this Report" (Copps 2009, 1). While rural connectivity was certainly mentioned in the NBP, it did not take front stage in the 400-page document. A distinct and dedicated conversation about *rural* broadband was lost in the shuffle as the FCC worked to fulfill Congress's order to draft the NBP.

Despite, or perhaps because of, its lackluster impact, *Bringing Broadband to Rural America* is a key document in the regulatory history of rural broadband because it was the first and only time the FCC seriously considered rural broadband as a unique policy issue.[5] What was clear then and remains clear today is that America requires a dedicated and coordinated policy approach to rural broadband deployment.

Meanings: The Discourse of Rural Broadband Policy

The next failure of rural broadband policy is meaning: What exactly is broadband? The FCC defines broadband as an "always-on" internet connection with a minimum download speed of 25 Mbps and upload speed of 3 Mbps. The 25/3 threshold was established in 2015, when the FCC increased it from 4/1; the threshold had been set at 200 kilobits per second (Kbps)/200

Kbps in 2010. Jameson Zimmer (2018) of BroadbandNow suggests that speeds were raised in 2015 not only to account for advances in technology, as the FCC claimed, but because this "was also in keeping with 2010 FCC Chairman Tom Wheeler's focus on drawing attention to the 'digital divide' in rural areas, as increasing the minimum speed results in less favorable coverage statistics." As such, it was both a practical and political decision to encourage providers to step up and connect the country as promised. That said, when this definition was drafted, 25/3 was on the verge of obsolescence; today it is painfully out of touch (Falcon 2020). The average download speed in the United States in 2018 was 96.25 Mbps and the average upload speed 32.88 Mbps, according to Ookla (2018). Moreover, the asymmetrical nature of the definition, which privileges download over upload, does a disservice to businesses that require equally fast upload speeds. Ernesto Falcon, senior legislative counsel of the Electronic Frontier Foundation, has called the 25/3 standard "both useless and harmful." He argues that

> it masks the rapid mobilization of high-speed access occurring in the United States and obscures the extent to which low-income neighborhoods and rural communities are being left behind. And, it attempts to mask the failures of our telecom policy to promote universal broadband. (Falcon 2020)

Using 25/3 allows the FCC to claim near-universal broadband in the country. It hides the unequal deployment of high-speed networks that privilege urban and wealthy communities and lets large telecommunication providers off the hook for failing to upgrade their networks. Moreover, in a time of social distancing and virtual learning, three college students sharing an apartment that is only able to access 25/3 would not be able to attend separate virtual classes held simultaneously—their apartment wouldn't have the bandwidth. Equally important to those college students may be the fact that Netflix recommends a minimum download speed of 25 Mbps to stream Ultra HD and 4K content. This means only one person could stream at a time. Jonathan Sallet (2019b) of the Benton Institute advocates for a minimum threshold of 100/100 for any provider seeking government funding. FCC commissioner Jessica Rosenworcel (2019) seems to agree, writing in a 2019 statement that "100 megabits per second is table stakes and we are going to need more symmetrical upload and download speeds as we move from an internet that is about consumption to one that is about creation."

While the FCC's current broadband definition fails to account for current and future usage, it does succeed at protecting "Big Telco" from expensive and resource-intensive upgrades to their networks. This is because 25/3 aligns with major telecommunications companies' choice of broadband technologies: DSL.[6] Recall from the introduction to this book that DSL uses the copper wires laid down or hung up decades ago to provide telephone service. They are the most prevalent broadband technology in rural America (if we disregard satellite). DSL networks, however, are slow, with a recent study pointing to a median speed of 10/1, far below the FCC's 25/3 definition (Gallardo and Whitacre 2019). Still, the FCC's definition of broadband is low enough for DSL networks to meet it, meaning that the companies managing these networks remain eligible for grants, loans, and regulatory favor.[7]

The FCC sees the technologies by which Americans access the internet—such as fiber, coaxial cable, DSL, fixed wireless, and satellite—as competitive and interchangeable (with the exception of cellular). This belief in interchangeability is based on a policy position of technological neutrality. Technological neutrality means that policies cannot discriminate against (or favor) a specific technology (Maxwell and Bourreau 2014). While intended to protect technological innovation, when coupled with a low-speed definition of broadband, this policy principle tends to favor incumbent telecommunication companies using DSL (e.g., Frontier, Windstream, CenturyLink), and satellite internet providers (e.g., Hughes, Viasat) over next generation technologies like fiber optics.

The discourse of technological neutrality permeates industry comments to the FCC in its investigations into rural broadband deployment. Telecommunication industries, associations, and even political entities like states want to ensure that any and all policy changes maintain a position of technological neutrality. The stakes are extremely high. If the FCC decides to endorse certain technologies over others (such as fiber), it would render the other technologies immediately obsolete and force companies employing them to change or get out of the business. The same thing would happen if the FCC were to raise the definition of broadband to a higher speed threshold (say 100/100 rather than 25/3). In short, it is in the vested interest of Big Telco to keep broadband policy technologically neutral and the threshold of broadband as low as possible to accommodate technologies like satellite and DSL.

When the FCC invited comment on its 2009 rural broadband strategy, numerous companies such as Sprint (cellular), Nextlink (fixed wireless), Hughes (satellite) and WildBlue (fixed wireless) advocated for technological neutrality. Hughes (2009), in fact, pointed out the FCC takes a technologically neutral perspective specifically because "satellite platforms may be inadvertently excluded by facially neutral requirements." The satellite industry found support in the NTCA—The Rural Broadband Association (2009), the trade association of the rural broadband industry:[8]

> If there were an economically feasible way that the most remote customers could be provided broadband through any method other than satellite, rural carriers would undoubtedly be doing so. Rural carriers currently use a variety of technologies to reach customers: DSL, fiber to the home/fiber to the curb, wireless (both licensed and unlicensed), satellite and cable modem. (13)

Ten years later, technological neutrality was the crux of UScellular's (2019) filings to the 2019 RDOF docket. The mobile provider argued for retention of the principle and a reduction in performance tiers "so that broadband providers planning to offer services below the gigabit speed—regardless of the technology they will use—have a fair opportunity to compete in the RDOF Phase I auction" (UScellular, 2019, 11). Filing in the same docket, Viasat (2019) added, "Excluding satellite providers through a high subtractive latency penalty would be disastrous for the Commission's efforts to expand access to supported broadband services, as well as inconsistent with the principles of technological and competitive neutrality" (3).

There is an assumption threaded throughout these comments that, as consulting group NRTC wrote in 2009, "great is the enemy of good." In other words, policymakers must be wary of making the standards for rural so high that it will depress investment. As the NRTC (2009) contended,

> Agencies should avoid any hard-line data speed standards and any "gold standard" level of service. Without question, the faster a service is the better. But in this case, great is the enemy of good. With millions of Americans lacking broadband, the goal should be to ensure access to best reasonable level of service, given all circumstances. . . . Consumers should not be forced to wait a longer period for a "gold" or "platinum" level of service that may never arrive. In areas where the delivery of any form of terrestrial broadband service is not feasible (or at least not in the near term), the agencies should acknowledge and support satellite delivered broadband, irrespective of speeds. (9)

These arguments were not lost on the FCC, and technological neutrality is found in the 2010 NBP, 2011 USF Order, and, years later, the 2019 RDOF Order. As the Commission noted in 2009,

> Every rural area presents its own special challenges, and a particular technological solution may be well-suited to one situation and poorly-suited to another.

Decision makers therefore should proceed on a technology-neutral basis—by considering the attributes of all potential technologies—in selecting the technology or technologies to be deployed in a particular rural area. (Copps 2009)

Invoking technological neutrality is a discursive tactic in telecommunications policy. More than that, when we place technological neutrality in conversation with claims to universal service, we see the creation of a discursive formation—a myth (Howarth 2000; T. van Dijk 1993).[9] There are four possible interpretations of this myth. First is the argument by satellite companies and DSL providers that technological neutrality is necessary so that they are not left out of funding decisions. Second is the argument from consumer interest groups that technological neutrality ensures that some form of internet connectivity—even satellite—is available to rural America. The third also comes from consumer interest groups and concerns community needs. More specifically, that each and every community should be able to decide for themselves which technologies best suit their needs. Fourth is the approach of legal scholar Jonathan Sallet (2019b), who advocates technological neutrality but only for those technologies that are truly economic substitutes and are scalable to meet future demands. As noted above, he advocates that public monies only fund networks that meet or exceed a 100/100 threshold. The State of Minnesota takes this approach with its Border-to-Border broadband program, which will fund any technology that is scalable to 100/100 or better. This by definition excludes both satellite and DSL from state-level funding.

On the surface, therefore, technological neutrality is neither a harmful policy nor an example of capture. It privileges community needs by not forcing every community to adopt a particular (and often expensive) technology, and it privileges innovation. It thus finds support with consumer and public interest groups, such as the Consumer Federation of America and Consumers Union (2009), who "urge[d] the Commission to adopt a least cost, technology neutral, no regrets approach to serving the un- and under-served in rural America and low-income inner city neighborhoods", and with small provider associations like the NTCA (2019), who cautioned against protecting networks that use antiquated technologies and are therefore "built to fail" in delivering next generation speeds. Policies should not be built around particular technologies, lest they risk undercutting superior technologies yet to be developed (Maxwell and Bourreau 2014). All petitioners agree on this point. When technological neutrality is paired with a slow speed threshold, such as the current standard of 25/3, however, it supports a politics of "good enough" by justifying the deployment and subsidization of outmoded technologies. Specifically, technological neutrality and a 25/3 definition keeps DSL and satellite providers at the table. This was the point made by ADTRAN in their 2019 RDOF filing:

> Technological neutrality does not mean that all access technologies should be subsidized regardless of their ability to support the broadband services and applications needed by consumers. Rather, it means that the ability of a given proposed service to meet the required performance should be evaluated without regard to the underlying access technology. (4–5)

Fast deployment and technological neutrality are present-minded policy goals that do not anticipate the growing consumption of data and the growing requirement of fast and symmetrical connectivity—what Sallet (2019a) calls "high performance broadband." Put differently, and to paraphrase the NTCA, policies may be technologically neutral, but they must not be "technologically blind."

Money: The Political Economy of Rural Broadband Policy

Because of the stringent belief in immediacy and neutrality, DSL and satellite-providing companies remain eligible for broadband grants, loans, and subsidies, which currently surpass $8 billion annually. The USF is the marquee vehicle for rural broadband subsidy. Originating in the 1996 Telecommunications Act as a subsidy for rural telephony, the 2010 NBP (FCC 2010) forcefully recommended the USF be restructured so as to subsidize broadband rather than, or in addition to, landline telephony. The aim was to transform the telephone-centric High-Cost Fund—that part of the USF dedicated to subsidizing companies serving high-cost (i.e., rural) areas—into the Connect America Fund (CAF), which is dedicated to broadband deployment.

The massive task of restructuring the USF began in 2011 and continues to this day. The USF is a cross-industry subsidy, wherein telecommunications companies pay into the fund and money is subsequently meted out to telecommunications companies to connect unserved areas and to schools, libraries, and hospitals to subsidize connectivity. Fees are levied only on the interstate telephone operations of telecommunications companies, including landline telephone service, cellphone service, and the vaguely worded broadband connections that are deemed "interstate." The fees are passed onto consumers through a line item on their monthly bills. In 2018, the

USAC allocated approximately $8.3 billion in subsidies to support broadband deployment across the country (not just to rural communities) and maintains four programs:[10]

1 Connect America Fund (formerly the High-Cost Fund) (subsidies for broadband providers serving high-cost areas): $5 billion
2 Lifeline (subsidies for low-income households that bring down the cost of telephone or broadband): $981 million
3 E-Rate (subsidies for schools and libraries that bring down the cost of broadband): $1.98 billion
4 Rural Health Care (subsidies for rural hospitals and health care facilitates that bring down the cost of broadband): $251 million (USAC 2019)

The CAF is the USAC's hallmark program and the primary vehicle for fixed rural broadband subsidization. As such, it will be the primary focus of this analysis. The CAF contains four subprograms:

• Connect America Phase II
• Connect America Phase II Auction
• Alternate Connect America Cost Model
• Mobility Fund II

Joining these are the recently announced $20.4 billion RDOF, which will replace the CAF Phase II (CAF II) program in 2021, and the $9 billion 5G Fund for Rural America, which will replace the long-defunct Mobility Fund II (not considered here).

The Connect America Fund Phase II   The CAF II began in 2015, when the USAC and FCC allocated $10 billion over six years (2015–2020) to what are known as "price-cap carriers."[11] Price-cap carriers are the largest telecommunications carriers—the legacy phone companies like AT&T, CenturyLink, Windstream, and Frontier—with a predominantly nationwide footprint. The FCC identified ten such providers, identified unserved and underserved rural areas, calculated a cost per location (between $52.50 and $198.60), and offered providers a set amount of money to provide service in these areas based on the cost to serve the subscriber (FCC 2014d, 2015b). Price-cap carriers were exclusively chosen because, at the time, it was reported 83 percent of rural unconnected peoples lived in price-cap territories (FCC 2011b). The decision to simply give price-cap carriers money, rather than requiring them compete for it through a competitive auction, however, was one rooted in both politics and history:

> We conclude that the Connect America Fund should ultimately rely on market-based mechanisms, such as competitive bidding, to ensure the most efficient and effective use of public resources. *However, the CAF is not created on a blank slate, but rather against the backdrop of a decades-old regulatory system.* The continued existence of legacy obligations, including state carrier of last resort obligations for telephone service, complicate the transition to competitive bidding. (FCC 2011b, para. 165, emphasis added)

Price-cap providers could pick and choose how much money they wanted to receive and which areas they wanted to serve. Verizon, for instance, refused most of its funding, and what it did accept it eventually turned over to Frontier after selling many of its copper lines to Frontier (Engebretson, 2015). Not surprisingly, price-cap carriers chose to serve the most populated areas, therein ensuring some return on investment. AT&T, for instance, only accepted funds for areas that surpassed what it had received in the CAF I (FCC 2015a). In the end, a total of $10.5 billion ($1.67 billion/year) was offered, and price-cap carriers accepted $9 billion or approximately $1.5 billion per year (table 2.1). As an example, the FCC offered Century-Link $514,334,045 per year to connect 1,190,016 locations. It eventually accepted $505,702,762 per year to serve 1,174,142 locations in thirty-three

Table 2.1
Recipients of Connect America Fund II

Company	Amount accepted (per year)	Total over 6 years (2015–2021)
CenturyLink	$505,702,762	$3,030,000,000
AT&T	$427,706,650	$2,560,000,000
Frontier	$283,401,855	$1,700,411,130
Windstream	$174,895,478	$1,040,000,000
Verizon	$48,554,986	$291,329,916
Fairpoint Communications	$37,430,669	$224,584,014
Consolidated Communications	$13,922,480	$83,534,880
Cincinnati Bell	$4,449,130	$26,694,780
Hawaiian Telecom Inc.	$4,424,319	$26,545,914
Micronesian Telecom	$2,627,177	$15,762,702
TOTALS	**$1,500,896,506**	**$9,005,379,040**

Source: FCC (2015a).

states, declining funding for locations in California, Mississippi, Oklahoma, and Wisconsin. In Minnesota, specifically, CenturyLink accepted $54 million per year to connect 114,739 households (Coleman 2018).

To receive CAF II funding, price-cap carriers must first be designated as eligible telecommunications carriers (ETCs) by the respective state utility commissions, which includes providing voice services (47 US Code §214(e)(1)). This is a holdover from the days of telephony and by definition excludes any company that does not intend to offer a voice package (FCC 2011b). Having been designated an ETC and accepted CAF II funding, price-cap carriers must build out broadband to 40 percent of funded locations by the end of 2017, 60 percent by 2018, 80 percent by 2019, and 100 percent by 2020.

Broadband for CAF II recipients was defined as 10/1, not the national standard of 25/3. This was raised from 4/1 in the initial 2011 proposal (FCC, 2011b), but it was not raised a second time, even after the FCC declared 25/3 to be the definition of broadband in 2015. The aim was for 4 million rural households to have 10/1 by the completion of the program (2020/2021). Reaching a 10/1 threshold is not an onerous condition for the largest providers. Technologically, this can be done via DSL or fixed wireless without much difficulty. In other words, the CAF II does not require providers to lay down fiber to the home, curb, or tower, and gives them no incentive to go beyond the 10/1 threshold minimum. As a result, communities receiving the CAF II and subsequently categorized as "served" may find themselves stuck in 10/1 or 25/3 purgatory for the better part of the 2020s, while urban America is moving to 100/100 and 1 Gbps/1 Gbps.

Even with these low requirements, certain price-cap carriers failed to meet buildout expectations. In January 2019, CenturyLink notified the FCC that it had failed to meet its deployment targets for eleven states: Minnesota, Colorado, Idaho, Kansas, Michigan, Missouri, Montana, Ohio, Oregon, Washington, and Wisconsin (Treacy 2019). The following January (2020) it again notified the FCC that it had failed to meet deployment targets, this time for twenty-three states. At the same time Frontier (2020) also admitted it had failed to meet benchmarks in thirteen states. Like CenturyLink, this was the second time in as many years that Frontier had failed to live up to its CAF II commitments (Brodkin 2020c). These admissions were not met with sanctions from the FCC.

Connect America Phase II Auction   In the original CAF II plan, price-cap carriers were offered $10.5 billion and accepted $9 billion. The remaining funds, or at least $1.98 billion, were put into an auction (Auction 903), where all eligible carriers (and not just price-cap carriers) could bid for this money to serve areas declined during the CAF II process. Eventually, $1.48 billion for ten years (2018–2028), or $148 million a year in annual support, was awarded to 103 bidders. These bidders cover 713,176 locations in forty-five states (FCC 2018a). According to the FCC, over 99.7 percent of these locations will receive at least 25 Mbps download speeds. Unlike the original CAF II program, which catered to telephone companies, all types of ISPs were eligible to bid on the auction, including satellite providers, cable operators, and electrical cooperatives. Indeed, the top four winners would all have been ineligible for telephone-based subsidies (see table 2.2).

One of the largest recipients of the auction was the Rural Electric Cooperative Consortium (RECC), an amalgam of twenty-one electrical co-ops, who came together to receive $182 million (over ten years) to serve 66,322 locations. The RECC was the third-largest winner and the "largest winner that pledged to build out service supporting gigabit speeds" (Engebretson 2018b). In contrast, the largest winner, AMG Technology (d/b/a Nextlink), a fixed wireless company which won $281 million, committed to simply providing "baseline" (25/3) or an unspecified "above baseline" (FCC 2018a). Viasat, a satellite provider, was the only bidder in the top four to offer only "baseline" (25/3) service. It won $122 million despite lackluster speeds and high latency (lag time between transmission and reception of data). Numerous critics have challenged the assertion that satellite technology can deliver broadband speeds (e.g., C. Mitchell and Trostle 2018), with Brian Whitacre et al.

Table 2.2
Top four winners of the CAF II reverse auction

Bidder	Technology	Amount per year	Commitment
AMG Technology (d/b/a Nextlink)	Fixed wireless	$28 million	Baseline (25/3) and above baseline
WISPER Inc.	Fixed wireless	$22 million	Above baseline
RECC	Fiber	$18 million	Gigabit
Viasat	Satellite	$12 million	Baseline (25.3)

Source: FCC (2018a).

(2018) observing that "satellite technology is highly susceptible to weather disruption; data latency is an issue; and data caps/costs are also concerning." Hughes, a competitor to Viasat, went so far as to argue that Viasat lied to the FCC about its connectivity potentials to be eligible for the reverse auction. Hughes (2019) virtue-signaled its decision not to compete in the reverse auction for this reason:

> Viasat participated in the CAF-II auction knowing the applicable technical requirements, and knowing that it could not meet them. In contrast, knowing those requirements, Hughes declined to participate. (1)

While the FCC eventually dismissed Hughes's complaint, that Viasat was not only permitted to compete in the auction, but came out successful, illustrates the pervasiveness of the politics of "good enough" and its concrete policy consequences. It reflects the FCC's historical propensity to favor incumbents' current rural networks, rather than to force them to upgrade.

The Alternate Connect America Cost Model   Smaller telecommunications companies, known as rate-of-return (RoR) carriers, have their own fund under the umbrella of Connect America. This is the Alternate Connect America Cost Model (A-CAM) fund. RoR carriers are the telephone companies serving rural and sparsely populated communities. They are not a legacy of the Bell system but rather independent telephone companies, mostly local, and oftentimes cooperatives.

When the CAF was first imagined, the FCC proposed $2 billion for the A-CAM. When the program took effect, however, that number was reduced to $1 billion a year for ten years (FCC 2018f). The program started in 2016 with USAC funding 175 companies for ten years using the allocated $1 billion/year (compare with the $1.5 billion a year divided by only ten companies over six years for price-cap carriers). In the Order that created the A-CAM, the FCC acknowledged the important role of local providers in achieving universal broadband:

> Rate-of-return carriers play a vital role in the high-cost universal service program. Many of them have made great strides in deploying 21st century networks in their service territories, in spite of the technological and marketplace challenges to serving some of the most rural and remote areas of the country. (FCC 2016, para. 2)

By most standards, RoR carriers are doing much more for less money and higher expectations than their price-cap cousins. All but the smallest A-CAM recipients (48.7 percent of locations) must guarantee 25/3 by the end of the subsidy. Telecommunications consultant Doug Dawson (2018a) has noted that while many price-cap carriers use their funds to "upgrad[e] to just enough speeds to get them over the 10/1 Mbps requirements, many small telcos are doing a lot more." This includes parlaying funds into seed money to build out fiber.

To incentivize more RoR carriers to deploy broadband at or above 25/3, the FCC announced the creation of a second round of A-CAM funding in 2018. Eligibility is met by being a RoR carrier and having not received A-CAM I funds (although the FCC did increase their funding as well). The FCC proposed to allocate $4.9 billion over ten years (2019–2028) for the A-CAM II. This will be divided between 171 carriers in thirty-nine states and tribal lands and translates into $200 per location served. The FCC also raised its expectations for the number of locations that receive 25/3 (Engebretson 2019b; FCC 2018f).

To be clear, both the CAF II and A-CAM programs exclusively fund telephone companies that offer broadband. Only the CAF II reverse auction opened funding to other companies, such as electric cooperatives, municipalities, and fixed wireless companies (see table 2.3).

There has been considerable critique of the CAF process, most notably because it replicates earlier telephone-era rules that favored incumbents (i.e., Bell companies). Here, the ten largest incumbents received $1.5 billion a year, while 175 RoR companies receive $1 billion a year and all others combined receive $148 million a year. Moreover, price-cap carriers are expected only to deliver 10/1 to their locations unless they promised a higher speed in the CAF II auction, while the national standard is 25/3— the standard required for RoR carriers. Many of the price-cap companies are not upgrading their copper wires to fiber but rather using DSL and fixed wireless to deliver 10/1 speeds (and often with a hefty subscription price tag) (Dawson 2018b). Here is Dawson's (2018c) critique:

> This disparity between rural haves and have nots is all due to FCC policy. The FCC decided to make funds available to rural telcos to upgrade to better broadband, but at the same time copped out and handed billions to the giant telcos to instead upgrade to 10 Mbps DSL or wireless. To make matters worse, it's becoming clear that AT&T and Verizon are intent in eventually tearing down rural copper, which will leave homes with poor cellular coverage without any connection to the outside world.

In contrast, the reverse auction received high praise from scholars and industry watchers for lowering the cost of service, attracting new providers,

Table 2.3
FCC/USAC rural broadband subsidy programs (2019)

	Amount	Expiration	Eligibility	Requirements	Critiques
CAF II	$1.5 billion/yr ($9 billion total)	2015–2021	Price-cap carriers (telephone)	100% 10/1 service to required areas by 2020	Favors large, incumbent, legacy telecommunications companies No incentive to upgrade connections Inaccurate broadband map
CAF II Auction	$148 million/yr ($1.48 billion total)	2018–2027 (app)	Designated ISPs in pre-determined geographic areas as determined by the FCC	Minimum of 10/1	Not enough funding Can be a 10/1 bidder or a gig bidder—whoever comes in under budget Inaccurate broadband map
A-CAM	$1.04 billion/yr ($10.4 billion total)	2016–2025 (app)	ROR carriers (telephone)	Scaled depending on size At least 50% directed toward 25/3	Not enough funding compared to CAF II Inaccurate broadband map
A-CAM II	$490 million/yr ($4.9 billion total)	2019–2028	ROR carriers who did not accept the A-CAM I	Scaled depending on customer size Emphasis on 25/3	Only for telephone companies Inaccurate broadband map
Mobility Fund II (defunct)	$453 million/yr ($4.53 billion total)	Unknown (2020–2029?)	Designated mobile providers in areas determined by the FCC	Upgrading connections to 4G LTE and 10 Mbps download	Relies on self-reported data from ISPs Inaccurate broadband map
RDOF	$20.4 billion (Phase I: $16 billion, Phase II: $4.4 billion)	2021–2030	Areas that are wholly unserved at 25/3 (Phase I) Areas that are partially unserved at 25/3 (Phase II)	Minimum of 25/3 but preference given to applicants offering gigabit speeds	ISPs that have received ReConnect or state funds are ineligible Still using 25/3 as baseline Inaccurate maps

and ushering in higher speeds (Glass and Tardiff 2019, 15). It was so highly regarded in fact that it became the basis for the FCC's newest RDOF, announced in the summer of 2019.

The decision to favor incumbent ISPs and telecommunications providers over a more democratic or horizontal subsidy architecture illustrates a trend within American communication policy that favors the status quo at the expense of alternatives and emerging actors (McChesney 1995). As McChesney (1995) writes, "The status quo, therefore, easily became the 'American Plan' where the public interest was assured, as it always was, by the machinations of the marketplace" (264). In this case, the marketplace is synonymous with the interests of the major telecommunications companies rather than the more diverse long tail of ISPs. Clearly, a more favorable route would have been for the billions allocated by the FCC for rural broadband for 2015–2020 to be distributed through a competitive auction that recognized all stakeholders and set minimum speeds at 25/3 (or higher!) rather than a ceiling of 10/1 for the large incumbents and a more demanding threshold for everyone else. Such a proposal, however, goes against the interest of major telecommunications companies, which have lobbied aggressively to maintain their incumbent status.

Mapping: The Critical Geography of Broadband Deployment

Just as the political economy of rural broadband subsidy is defined by the power of incumbent telecommunications companies to set their own terms, so too is the mapping and measurement of broadband deployment. To determine what areas of the country are "served" and "underserved" by fixed broadband providers, the FCC collects deployment data from ISPs through a document known as Form 477. ISPs fill out and submit Form 477 twice a year, and the data is used to inform the FCC's annual *Broadband Deployment Report*,[12] which is mandated by Section 706 of the 1996 Telecommunications Act. The information from Form 477 is also used to populate the FCC's National Broadband Map, a searchable online map the FCC inherited from the NTIA in 2017.[13]

As noted in the introduction to this book, the FCC's *2020 Broadband Deployment Report* found that 22.3 percent of rural Americans lack access to a fixed home broadband connection of 25/3 (FCC 2020a). This was based on data collected from 2018 (there is typically an eighteen-month lag between collection and publication). According to the FCC (2020a), the percentage of unconnected rural Americans shrank from 32.3 percent to 22.3 percent between 2016 and 2018. The FCC crowed over this reduction in the unconnected, proclaiming, "The digital divide has narrowed substantially, and more Americans than ever before have access to high-speed broadband" (2). The problem is that these percentages are woefully inaccurate, made so because of fatal flaws in Form 477, flaws that serve the interests of ISPs.

There are three major structural flaws with Form 477 (GAO 2018). The first is the granularity of the data collected. The data is reported by census block, rather than by residence or business. A census block is the smallest geographic area used by the Census Bureau. There are 11,166,336 census blocks in the country. In a city, a census block is usually a city block. But in rural areas they can be substantially larger. The largest census block in the country is in Alaska and measures over 8,500 square miles (FCC 2020d). Since ISPs report data by the census block rather than the individual household, a census block is considered entirely served so long as at least one building (house or business) receives broadband. I live in census block 515400006001012, and CenturyLink is the largest provider in my area. As long as one house in my census block has broadband, CenturyLink can claim on Form 477 that the entire census block is served with broadband, even though the individual experience of users and households may vary. Of course, it is in the interests of CenturyLink, or any ISP, to report complete coverage of an area and to claim dominant status in a market. Form 477 helps them do this.

This lack of granularity engendered by Form 477 means that the FCC has grossly overestimated how much of the country—rural or urban—has access to broadband. A recent study by Sascha Meinrath (2019) for the Center for Rural Pennsylvania examined the state of broadband connectivity in Pennsylvania. The FCC says Pennsylvania is 100 percent served with speeds of 25/3. In contrast, the study found that there is not a single county where at least 50 percent of the population receives broadband as defined by the FCC. This deficit was corroborated in a 2020 report by Broadband-Now, which conducted a study of 11,000 addresses across the country and concluded that 42 million Americans (or double the FCC's 2019 estimate) "do not have the ability to purchase broadband internet" (Busby and Tanberk 2020). Both studies suggest the FCC's map is off by at least 50 percent. Microsoft wholeheartedly agrees with this conclusion and goes even further. In a report issued late in 2018, Microsoft estimated that 162 million people, or 50 percent of the US population, do not have access to broadband speeds

(Kahan 2019). Microsoft came to this number by evaluating the millions of Americans who download Office updates. To be sure, this number attests that at least 162 million Americans have *internet*—it is a measure of speed, not of availability—but it nevertheless departs dramatically from the FCC's data, which places the number of Americans without broadband speeds at around 19 million in 2020 (21 million in 2019).

The second structural flaw in Form 477 is the data collection process. ISPs self-report the data, and the FCC does not conduct its own audits. To make matters worse, ISPs are only required to report *advertised* rather than *actual* speeds. Worse still, when an ISP reports an area as "served," it does not mean that it has actual service, only that the ISP can connect the area in under ten days. This gives the impression that communities are receiving much faster speeds than what consumers are actually experiencing. For example, the FCC says that CenturyLink provides 100 Mbps download service to Crozet, Virginia, while the actual median download speed is only 7.9 Mbps (M-Lab 2020). ISPs are also not required to report prices, thus making it impossible to compare service or evaluate accessibility. Can broadband be considered "accessible," for instance, when a basic connection is priced over $100/month?

The third structural flaw in Form 477 recalls the myth of technological neutrality. Here, the FCC considers satellite as "fixed" broadband and therefore interchangeable with cable, DSL, and fiber. A connection is a connection is a connection, according to the FCC. As we learned in the introduction to this book, however, satellite cannot possibly measure up to a wired connection as it is beleaguered by high latency, low data caps, intermittent service, and high prices. Suggesting that satellite is a complementary fixed connection gives an inflated impression of broadband deployment. By including satellite as a broadband option, however, the FCC can lay claim to more Americans connected to the digital grid (Cooper 2018a). The issue for rural America is that wired providers may avoid serving these certain communities because they are already "served" by a satellite provider. As Stephen Cobb (2011) noted,

> At the federal government site broadband.gov, run by the FCC, you can see Satellite listed as a type of broadband, despite the fact that the two main providers of such service avoid using the word "broadband" when they are pitching their service. So why include satellite alongside DSL, cable, wireless, and fiber? The answer may lie in pro-satellite lobbying. The logic for such lobbying is simple: If it can be said that satellite is a broadband option for rural communities, as listed by the FCC, then terrestrial telcos can argue there is no compelling need to provide those communities with alternatives.

90   Chapter 2

The implications of these structural flaws in Form 477 and the FCC's broadband mapping methodology are tremendous for rural communities. First and foremost, Americans are given the mistaken impression that the country is well served by broadband providers and that competition is healthy (see figure 2.1). Even more egregious are the implications for future funding. As noted above, when an area is considered "served" by a broadband provider, that area is ineligible for future FCC subsidy.[14] Thankfully, the FCC does not take satellite into account when deciding on areas eligible for USF programs (see FCC 2020f, para. 13). If it did, the majority of the country would be considered served because satellite providers cover 99 percent of the United States. Nevertheless, communities are being overcounted and are therefore missing out on funding opportunities. As the previous section detailed, there is at least $5 billion in funding available for rural broadband deployment, but dozens if not hundreds of communities

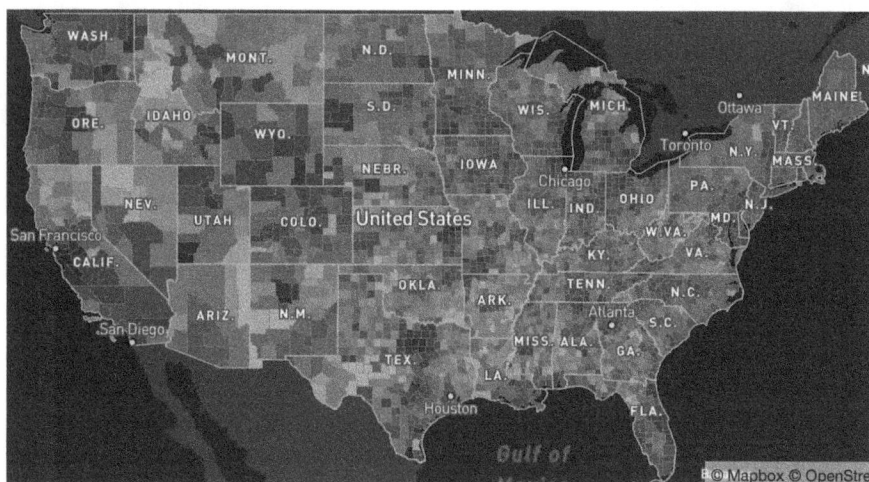

Number of Fixed Residential Broadband Providers

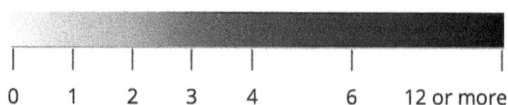

0   1   2   3   4   6   12 or more

Figure 2.1
The National Broadband Map as of 2018.
Source: FCC (2018e).

are deemed ineligible for these dollars because of the inaccurate data collected by Form 477.

Louisa County, Virginia, exemplifies this calamity. Louisa is a rural county with a population of 36,778 spread across 511 square miles. According to the FCC (2019c), Louisa is 100-percent served with broadband provision of at least 25/3 (see figure 2.2).

When satellite is removed, the level of connectivity drops by almost 40 percent (see figure 2.3).

When we drill down even further, crowdsourced speed tests conducted by M-Lab report an average down speed in the county of 3.91 Mbps and average upload speed of 1.69 Mbps (figure 2.4). These averages did not meet the broadband standards of 2015, let alone 2020. Despite these errors the FCC considers Louisa fully served. As noted above, because the FCC does not take satellite into account for USF awards, portions of Louisa remained eligible for the first phase of the RDOF (FCC 2020b).

There are important political economic reasons for why the FCC has little interest in changing the provisions of the map, aside from nominal adjustments. As Karl Bode (2018b) reported in a scathing critique of the broadband map for *The Verge*, ISPs "are heavily incentivized to overstate speed and availability to downplay industry failures." This echoes the language of critical cartographers, who note "behind the map-maker lies a set of power relations, creating its own specification" (Harley 1988, 287). Keeping the map as is allows the FCC to claim, as it did in the 2020 *Broadband Deployment Report*, that America is being well served by ISPs, enabling it to sidestep the need to take regulatory action. In the end, not only have the FCC and ISPs taken credit for closing the digital divide where no credit is due, but they have condemned hundreds of communities to broadband obscurity.

The Case of the Rural Digital Opportunity Fund

When Ajit Pai became chair of the FCC in January 2017, one of his priorities was the closing of the digital divide. Touting his upbringing in rural Kansas, Pai (2017a) wrote in the *Wichita Eagle* in September of that year:

> I know the value of growing up in a rural community and want to ensure future generations will be able to have that same experience. That's why, as Federal Communications Commission chairman, I've made it my No. 1 priority to close the digital divide. . . . In my view, every American deserves digital opportunity.

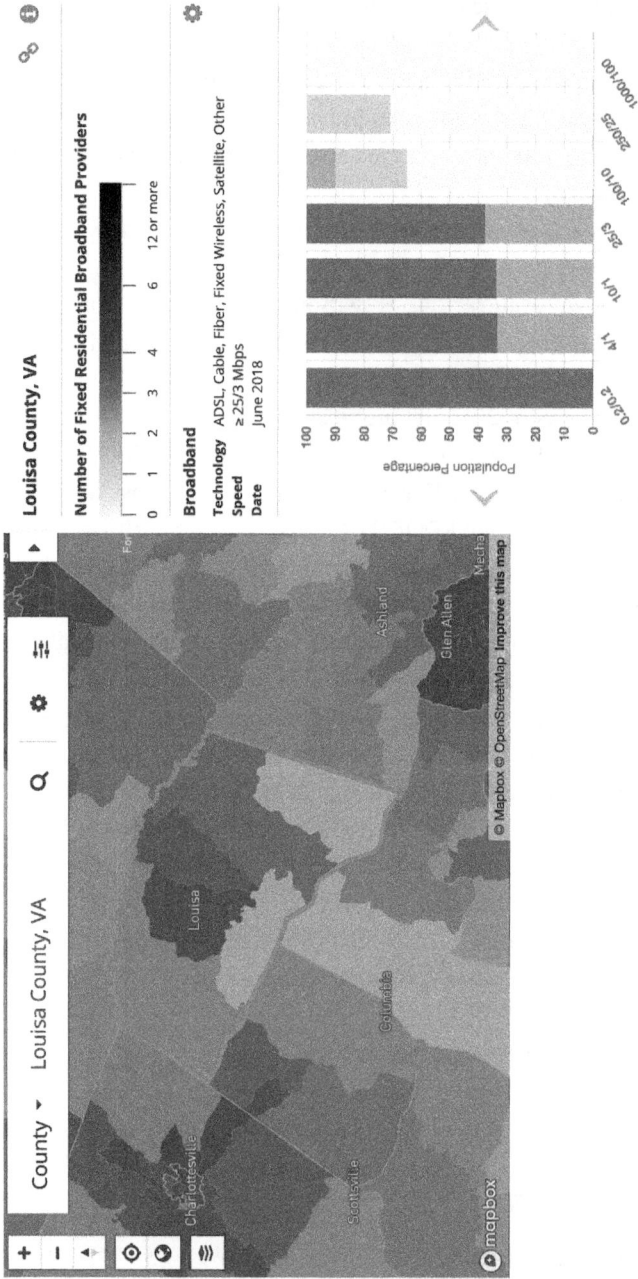

Figure 2.2
Broadband deployment in Louisa County, Virginia.
Source: FCC (2019c).

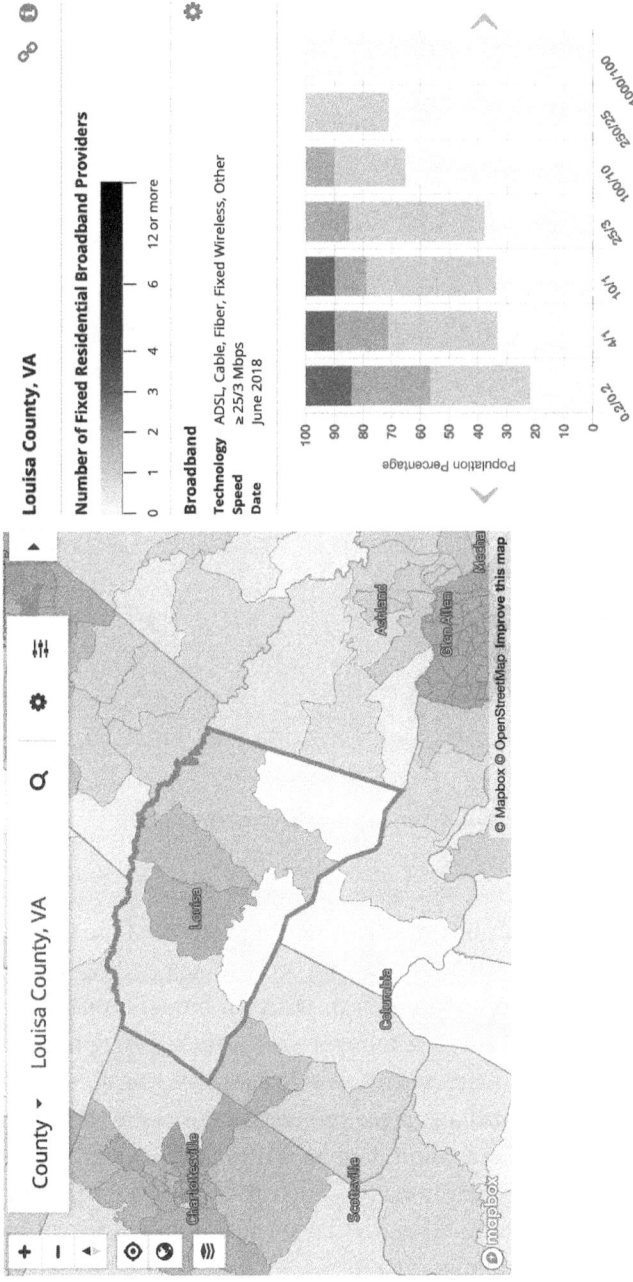

Figure 2.3
Broadband deployment in Louisa County, Virginia (without satellite).
Source: FCC (2019c).

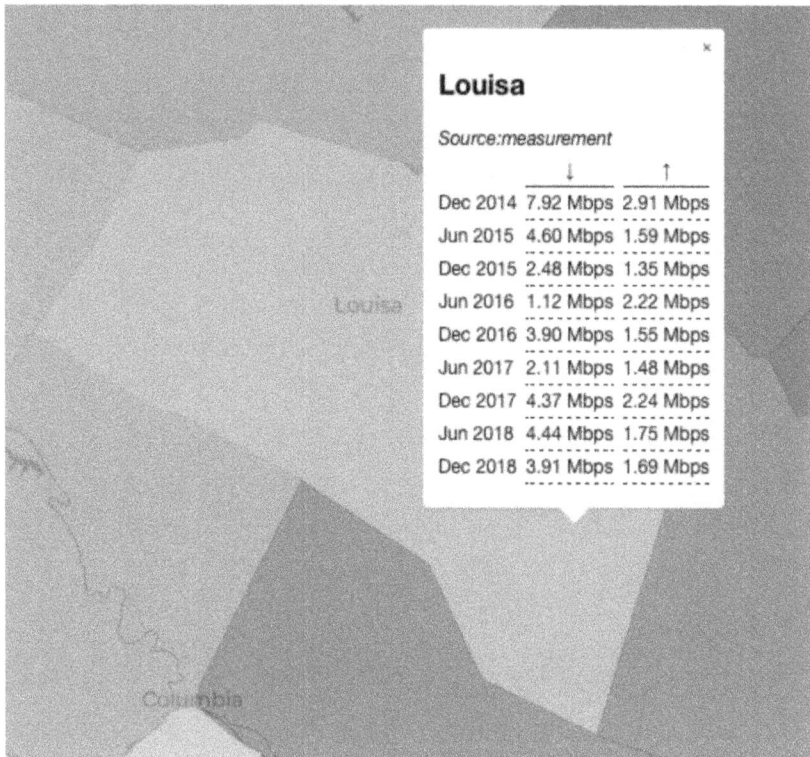

Figure 2.4
Average download and uploads speeds from Louisa County, Virginia.
Source: M-Lab (2019).

Pai traveled the country with this message, meeting with rural broadband providers, farmers, and rural senators, and even declared August 2017 "Rural Broadband Month" (FCC, n.d.). With these meetings he brought with him a deeply entrenched neoliberal ethos, claiming it was the rules and regulations of previous FCCs that "deny many Americans high-speed Internet access and jobs" (Pai 2017b). Pai used this argument to justify the repeal of net neutrality in December 2017 (Pickard and Berman 2019). It also served as the justification for other pro-industry initiatives, such as repealing the mandate that telephone providers ensure customers have equal service before they abandon their copper wires (Brodkin 2017b).

In a seeming about-face to his market fundamentalism, however, Pai's FCC announced the creation of the RDOF in April 2019. This included a whopping $20.4 billion to be spent over ten years to bridge the rural-urban digital divide. The FCC called the RDOF "a critical next step in the FCC's ongoing effort [to] provide rural America with the same opportunities available in urban areas" (FCC 2019b). To this, Pai (2019b) added:

> In short, we're proposing to connect more Americans to faster broadband networks than any other universal service program has done. I'm excited about what this initiative will mean for rural Americans who need broadband to start a business, educate a child, grow crops, raise livestock, get access to telehealth, and do all the other things that the online world allows.

While a massive financial commitment to deployment, the RDOF architecture seems to replicate many of the existing issues inherent within broadband policy already discussed in this chapter. Indeed, its structure underscores the endemic politics of "good enough" and reaffirms Pai's commitment to industry profit and favor.

The RDOF will replace the CAF II program, which expires in 2021. The $20.4 billion pledged by the FCC will be spent in two phases. The first phase will mete out $16 billion starting in 2021 to connect completely unserved census blocks. In the second phase, the remaining $4.4 billion will be awarded to partially served census blocks. As with the CAF II auction, the funds will be spent over the course of a decade. The minimum speed threshold for award winners is 25/3—the same speeds that have defined broadband policy since 2015. That said, the FCC will favor applicants who pledge to provide above-baseline higher speeds (100/20 with an allowance to use at least 2 terabytes [TB] of data per month) and gigabit speeds (1 Gbps/500 Mbps) with a 2 TB cap (FCC 2020f).

From the outset, it is important to note that the RDOF is not a new program, as expressed by the FCC, but rather a rearticulation of the CAF II. To its credit, the RDOF will take the form of a reverse auction, mimicking the CAF II auction by awarding funds to the lowest bidders (amid other criteria such as speed and latency) for predesignated locations. Rural broadband consulting group Conexon took note of these changes from the original CAF II program, lauding the RDOF as "a true opportunity for rural America," adding "it is also an opportunity for the Commission to begin to redeem itself for past mistakes" (see Chambers 2019).

Despite garnering support, the RDOF faces significant criticism. First and foremost, the FCC will rely on the flawed maps and data collection described in the mapping section of this chapter. This occurs in spite of Congress passing the Broadband DATA Act in March 2020, ordering the FCC to revise its methodology. In response, the FCC cried poor, arguing that until Congress allocates funding for the change, the Commission's hands are tied (Eggerton 2020).[15] Relying on the flawed maps mean that many un- and underconnected communities will be ineligible for funding because they are erroneously deemed "served" on the map. As FCC commissioner Jessica Rosenworcel (2020) stated in a press release following a meeting on the RDOF, "Haste makes waste: this effort has been pushed out so fast I fear we are only starting to understand what is not workable in this framework."

Second, the baseline speed threshold—25/3—has been criticized for failing to encourage providers to deploy faster networks. Here, we see the FCC reaffirm its commitment to the politics of "good enough" when it justified the 25/3 baseline:

> Although we have a preference for higher speeds, we recognize that some sparsely populated areas of the country are extremely costly to serve and providers offering only 25/3 Mbps may be the only viable alternative in the near term. Accordingly, we decline to raise the required speeds in the Minimum tier and we are not persuaded that bidders proposing 25/3 Mbps should be required to build out more quickly or have their support term reduced by half. (FCC 2020f, 19)

In other words, the FCC is comfortable seeing rural communities connected with 25/3—and then made ineligible for all future grants since they will now be "served" despite an average download speed in the United States of 96.25 Mbps (Ookla 2019).

Eligibility is a third major issue. The FCC opened up eligibility to any and all broadband providers, unlike the CAF and A-CAM programs which specifically target telephone companies. Large telecommunications companies, such as Frontier and CenturyLink, will continue to be eligible for the RDOF despite failing to live up to their commitments in the original CAF II program. In the fall of 2020, the FCC announced a list of companies who had filled out short-form applications. This list included those telecommunications providers mentioned above; other telecommunication companies, such as Windstream; some of the largest cable companies, including Charter, Cox, and Altice; satellite providers Hughes and Viasat; and mobile providers, such as Verizon and UScellular. AT&T, T-Mobile, and Comcast all announced they would not compete for RDOF funds (Dano 2020; FCC 2020g). Fixed wireless providers such as Rise Broadband and Starry, and rural cooperatives, including the Farmers Mutual Cooperative, Federated Telephone Cooperative, and Paul Bunyan Rural Telephone Cooperative, also filed paperwork with the FCC indicating their intentions to apply (FCC 2020g).

Earlier in 2020, Frontier Communications petitioned the FCC to exempt 16,987 locations across the country from RDOF eligibility because it claimed it had already deployed 25/3 to those areas. Given Frontier's 2019 admission that it had not fulfilled its yearly CAF-II buildout requirement, as well as its reliance on DSL networks and its April 2020 bankruptcy, this request was met with intense skepticism. These points were not lost on WISPA and the NRECA:

> We find it difficult to believe that Frontier was able to provide voice and 25/3 Mbps service in each of these 16,000 census blocks in just eight months, and question how this is possible, especially in light of Frontier's operational issues and financial woes that led to its filing of a bankruptcy petition (quoted in Engebretson 2020b; see also Brodkin 2020e)

Similarly, Charter Communications (d/b/a Spectrum) asked the FCC to exclude 2,127 census blocks in New York State that it had promised to connect as part of the conditions of its merger with Time Warner Cable (Brodkin 2020d). The FCC upheld Charter's request but denied Frontier's, leading *Ars Technica* to write the headline "FCC helps Charter avoid broadband competition" (Brodkin 2020d). Last in this list of eligibility issues was a summer 2020 Order by the FCC that low-Earth orbital (LEO) satellite broadband providers—an untested broadband service made famous by Elon Musk's company SpaceX and its ISP Starlink—would be eligible to compete in the RDOF. Originally, the FCC had expressed "serious doubts" that LEO-based networks, which have yet to be deployed on a nationwide basis, could provide gigabit speeds, thus classifying them as "high-latency providers" (meaning they would begin the auction at a deficit). The summer 2020 Order reversed course and allowed them to compete in the low-latency tier so long as they can prove they can deliver latencies below 100 ms (Brodkin 2020a). With these two examples in mind—Charter and SpaceX—there is reason to be concerned that the RDOF will end up favoring the largest providers, who can afford to undercut smaller providers and whose technology—in the case of SpaceX—is unproven at scale.

Worse than the shenanigans of these giant providers are the eligibility criteria themselves. The FCC declared that any location that has received either state broadband support *or* funding from the RUS's ReConnect Program (see chapter 3) is ineligible for the RDOF. This eliminates a plurality of rural communities and counties and actively punishes those states that have designed their own broadband programs. According to the Pew

Charitable Trusts (2019), twenty-nine states have dedicated broadband funding programs. These states include Minnesota (profiled in chapter 4); Illinois, which, in 2019 pledged $420 million for broadband deployment; and New York, which has pledged $500 million. The RDOF effectively takes the legs out from under these state programs.

In addition to eligibility, public-interest organization Public Knowledge has questioned the FCC's legal authority to fund broadband through its USF programs since the FCC disclaimed the ability to oversee ISPs in its 2018 repeal of network neutrality (FCC 2018c Stella 2019). Harold Feld, senior vice president at Public Knowledge, stated that while any new funding deserves applause, "this new proposal deserves, at best, a Nancy Pelosi applause rather than the standing ovation Chairman Pai believes he deserves" (qtd. in Stella 2019).[16]

There are reasons to celebrate the RDOF: when it begins to provide funds, the RDOF will be the largest federal intervention ever into broadband. That said, there are even more reasons to be worried that history will repeat itself because the RDOF's policy architecture continues to replicate the politics of "good enough" described throughout this chapter.

Conclusion: The Politics of "Good Enough"

When read in concert, the four failures of rural broadband policy—management, meaning, money, and mapping—underscore a recurrent and dominant politics of "good enough." According to the FCC, for rural communities 10/1 and 25/3 are "good enough," DSL is "good enough," satellite is "good enough," and census blocks are "good enough." This belief serves the material interests of incumbent telecommunications companies by justifying their broadband deployment strategies while simultaneously disenfranchising rural communities and stymieing efforts to close the digital divide.

This political discourse is captured by the phrase "good is the enemy of great." This is the reverse of the expression "great is the enemy of good," which refers to the idea we that should not let desires for perfection obfuscate implementation. The latter phrase is attributed to Voltaire (1772), who wrote in *La Bégueule* "the best is the enemy of the good." While this is sage advice that reminds us of the importance of compromise, in relation to rural broadband the politics of "good enough" is invoked to justify the deployment of subpar technologies, massive subsidies to incumbent providers, and a lack of accountability. In other words, when it comes to rural broadband "good" is not "good enough." Nevertheless, "good enough" is the dominant ontology of American rural broadband policy. We must demand greatness.

Rural America has long suffered from what Marc Raboy and David Taras (2004) call a "politics of neglect" when it comes to development. Indeed, a 2015 *New Yorker* piece categorized many rural communities as "left behind places" (Hendrickson, Muro, and Galston 2015). The past decade has seen telecommunications policy join this politics, both programmatically and discursively. The idea that broadband is "good enough" for rural Americans parallels demands that rural residents simply pack up and move to cities if they want to enjoy the trappings of modern life (see Jones 2018). As Libertarian commentator Nick Gillespie has argued, "The answer to people being 'left behind' isn't to bring the future to them (especially through tax dollars, which farmers and rural states soak up at massive rates). It's to make it easier for them to move" (qtd. in Jones 2018). Such neoliberal discourses belie the material realities of rural America and undercut the rights of Americans to live where they choose. The politics of "good enough" endemic within rural broadband policy joins these neoliberal calls of personal blame by suggesting that if rural Americans want faster broadband, they can switch to (a nonexistent) competitor, pay more (for nonexistent service), or move. More importantly, it denies young people in these communities the opportunities to thrive in the digital ecosystem of their peers.

That rural broadband is "good enough," or should not be "gold plated" as telecommunications and satellite companies contend, demonstrates how these actors prefer a digital divide to universal service because they receive the bulk of subsidies. More specifically, it demonstrates how major telecommunications companies continue to receive FCC favor despite lackluster network deployment *and* remain eligible for the $5 billion allocated annually by USAC to subsidize rural broadband deployment.

With the commencement of the RDOF in 2021, it has never been more important to expose and understand the politics of rural connectivity. As the FCC seeks applications for the first phase of the fund, it is vital that the mistakes of the past—documented in this chapter—are not replicated. Large telecommunications companies have failed to connect the country, but the FCC's uncompromising belief in the beneficence of the private market has blinded it to this reality. To connect the unconnected, the United States must reinvent broadband policy from the grassroots, beginning and ending with what matters most: people.

3   The Rural Utilities Service: The Reluctant Regulator

In 2009 the United States was in the grips of the worst recession in almost a century.[1] The gross domestic product had fallen by 4 percent, representing a $600 billion loss. Ten million homes were foreclosed, and 8.7 million Americans lost their jobs. Rural America, already economically vulnerable, was particularly impacted (Bennett et al. 2018; Lowrey 2017).

Amidst this political economic turmoil, Congress passed the 2009 American Recovery and Reinvestment Act (ARRA), an $831 billion stimulus package to jump-start the economy. Not since Franklin Roosevelt's New Deal had there been such a massive investment of public money into the national economy.[2] Significantly, ARRA included $7.2 billion for rural broadband deployment. This was, and remains, the largest single investment in telecommunications by the federal government.[3] In a follow-up hearing on the broadband portion of the stimulus package, Senator John Rockefeller of West Virginia explained the importance of broadband to the country:

> Without broadband, we won't have an economy. With the networks that we could produce, if we would, we can change education, we can change the way people think about the way they look at the world, the way they look at each other, interracial matters, the world at large. They're all available if we have broadband; and if we use it properly, we can learn how to be civil to each other, which would be a shock, but which would be very good for America. (United States Congress 2009, 1)

ARRA funds were split between two agencies: the RUS of the USDA and the NTIA. That part of this capital injection went to USDA recalls the history of rural connectivity from chapter 1. In the 1930s it was the USDA (through the REA) that was given the monumental task of connecting rural America to the electrical grid. In the 1950s, the country again looked to the USDA and not the FCC, which had been created in 1936, or AT&T, the great octopus of telecommunications, to spearhead the deployment of rural telephony. In 2009, the USDA would yet again be called upon to connect the unconnected.[4]

This chapter explains the USDA's role in rural broadband deployment and regulation. That the USDA has (some) power over broadband deployment presents a challenge for policymaking. Specifically, it raises the question: Who should be in control of the nation's rural broadband strategy, the USDA or the FCC? This chapter contemplates the RUS's legitimacy to craft rural broadband policy and its authority to carry it through to fruition. This is accomplished by an analysis of the broadband funding programs of the RUS, along with an analysis of reports and reviews of these programs and interviews with key stakeholders.

I make three arguments in this chapter. First, rural broadband meets the definition of what is called a polycentric regulatory regime (Baldwin et al. 2012; Black 2008). That is to say, authority over rural broadband policy is split between multiple regulators. Polycentric regulation comes with its own set of problems, particularly around the concept of regulator legitimacy. One of the key questions I ask and answer in this chapter, therefore, is: What is the RUS's legitimacy when it comes to crafting rural broadband policy? My second argument is that the RUS represents a "reluctant regulator." By this, I mean that the RUS flirts with being a regulator at certain times and a policymaker more broadly and a passive stakeholder at other times. The RUS is a stakeholder and actor in rural broadband with ability to influence policymaking at the highest level. It even has regulatory responsibilities vis-à-vis rural broadband in terms of funding, speed requirements, and the definition of "rural." But through its own communication and the communication of outside actors, its role in rural broadband policy is ultimately unclear: regulator, policymaker, banker, champion—sometimes all of these at the same time. Such uncertainty leads to regulatory confusion and ineffective policy decisions, most notably regarding grants and loans—the key components of the RUS's regulatory toolkit. My third and final argument is that the RUS—with its role clarified and supported both internally at the USDA and externally by stakeholders and Congress—is a crucial player in the future of rural broadband.

This chapter begins with a brief discussion of the concept of legitimacy in policymaking and regulation and an overview of the RUS's current broadband programs. The remainder of the chapter is devoted to primary research and is modeled after Julia Black's (2008) three roads to legitimacy:

(i) the institutional embeddedness of regulators, be they at the national, sub-national, supranational, or global level and the role of that institutional environment in the

construction and contestation of legitimacy; (ii) the dialectical nature of accountability relationships; and (iii) the communicative structures in which legitimacy
claims and accountability relationships are articulated and constituted. (144)

In the first section detailing my research, I describe the RUS's legislative
authority. In the second section, I outline critiques of the RUS, exploring its
accountability structures and focusing on performance, eligibility standards,
overbuilding, and definitions. Next, I examine its communicative structures
and "narratives of legitimacy" (M. Price 2012). Through these three sections
we see the emergence of an identity crisis at the RUS regarding rural broadband. The RUS sees itself first and foremost as the champion of rural America
on the federal stage. Legislatively, it is both a bank (in providing broadband
grants and loans) and a policymaker (in crafting the eligibility requirements
for said grants and loans). At times, however, Congress also imposes on the
RUS the roles of regulator, researcher, and coordinator—a dizzying array of
titles and responsibilities for an erstwhile understaffed and underresourced
agency. The penultimate section offers a case study on the 2019 ReConnect
Program. Here, we see the repetition of critiques and legitimacy challenges
that have long beleaguered the RUS. I conclude by pointing a direction forward for the RUS and outlining the need for it to spearhead the nation's
rural broadband deployment plan.

The Problem of "Many Hands"

What happens when multiple regulators share policy responsibility? Legal
scholars call this situation a "polycentric regime" and highlight the compet-
ing claims to legitimacy that may occur between regulators (Baldwin et al.
2012; Black 2008). Said differently, with multiple regulators and authority
figures, policymaking runs into a problem of "too many hands." As Black
(2008) notes,

> Polycentric regimes at any level (sub-national, national, supranational, global)
> pose the problem of "many hands." . . . The different regulatory roles and respon-
> sibilities of identifying goals, formulating standards, monitoring and enforcement
> are often dispersed between a number of participants, with significant implica-
> tions for accountability. (143)

The legitimacy of regulators, which Manuel Puppis and I defined in a 2018
paper as "rooted in the acceptance of an organization by others," may be
diminished by polycentric regulation, since those regulated find themselves

answering to multiple masters (Ali and Puppis 2018, 277). Nevertheless, legitimacy is a vital component of policymaking:

> In a governance or regulatory context, a statement that a regulator is "legitimate" means that it is perceived as having a right to govern both by those it seeks to govern and those on behalf of whom it purports to govern. (Black 2008, 144)

Outside actors are therefore key to establishing legitimacy:

> Legitimacy rests on the acceptability and credibility of the organization to those it seeks to govern. Organizations (regulators) may claim legitimacy, and may perform actions and enter into relationships in order to gain it. But legitimacy is rooted in the acceptance of that organization by others, and more particularly in the reasons for that acceptance. (Black 2008, 144)

Legitimacy, therefore, requires buy-in from both the regulator itself and outside stakeholders. This requires what Monroe Price calls (2012) "narratives of legitimacy," the stories and appeals an actor makes to justify their actions. "The shaping of these narratives," Price contends, "is a product of the discursive environments in which the state is understood, influenced by everyday speech and strategic communications" (29). While Price thinks of these as narratives that justify the existence of the state, I use the term to think through the narratives that the RUS and other actors use to justify or challenge its actions.

RUS Programs

The RUS derives its authority over rural telecommunications from the 1936 Rural Electrification Act (see chapter 1) and the various incarnations of the Farm Bill, which periodically modifies the act to include new technologies like broadband (which happened in 2002).[5] The RUS's primary mandate is to provide private companies—most notably cooperatives—with financial incentives to connect rural America. These incentives are derived from loans and grants. In other words, the RUS exists to correct for market failure.

While the RUS has required telecommunications loan and grant holders to build their networks with internet capability since 1995, the first pilot program specifically targeting household broadband began in late 2000, when $100 million in loans was made available "to finance the construction and installation of broadband telecommunications services in rural America" (RUS 2000a, 75920). The pilot was continued in 2001–2002 with $80 million in loans made available by Congress.[6] The program was deemed a success and made permanent through Section 601 of the 2002 Farm Bill.[7] Here, Congress determined that the RUS would be the agency for rural broadband because of its previous accomplishments in electricity and telephony (Romm 2015).

Since 2000, the RUS's broadband portfolio has grown significantly and now consists of four permanent programs and almost $800 million in annual loans and grants (see table 3.1). In 2018, it was allocated an additional $600 million from the Consolidated Appropriations Act to pilot a ReConnect Program geared toward connecting the most unserved and underserved rural areas. In total, the RUS controls roughly $1.4 billion in rural broadband loans and grants. Since 2004, 704 broadband projects have been funded with $8.6 billion in loans and $144.8 million in grants (Kruger 2019). While not an insignificant amount of money, the RUS's rural broadband commitments pale in comparison to the total loan and grant budget of the Office of Rural Development, which in 2017 stood at $8.8 billion, and the budget of USDA programs at $137 billion (USDA 2018a).

The Telecommunications Infrastructure Loan Program (see table 3.1) is the RUS's longest-serving telecommunications program, dating back to the telephone program of 1949. The Rural Broadband Access Loan and Loan Guarantee Program is the marquee vehicle for rural broadband deployment. As of 2019, it offered $29.8 million in loans through an appropriation of $5.8 million. According to the CRS,

> Entities eligible to receive loans include corporations, limited liability companies, cooperative or mutual organizations, Indian tribes or tribal organizations, and state or local governments. Eligible areas for funding must be completely contained within a rural area (or composed of multiple rural areas). Additionally, at least 15% of the households in the proposed funded service areas must be unserved, no part of the proposed service area can have three or more incumbent service providers, and no part of the proposed service area can overlap with the service area of current RUS borrowers or of grantees that were funded by RUS. (Kruger 2019, 5)

The minimum loan is $300,000, while the maximum is $25 million. The RUS established a minimum speed requirement of 25 Mbps download and 3 Mbps upload, in line with the current definition of broadband set by the FCC (see chapter 2). You will notice in table 3.1, however, that buildout requirements for the different broadband initiatives fluctuate between 10/1 and 25/3 depending on the program. Speed thresholds are one of the more unambiguous policymaking abilities of the RUS (through the Rural Electrification Act).

Table 3.1
RUS broadband loan and grant programs

Program name	Date	Authorizing legislation	Budget	Aim	Eligibility	Speed
Rural Broadband Access Loan and Loan Guarantee ("Farm Bill Broadband Loans")	2000 (made permanent in 2002)	Section 6103 of the Farm Security and Rural Investment Act of 2002 (amended the Rural Electrification Act of 1936)	$5.8 million, which allows for loans totaling $29.8 million	Construction, improvement, or acquisition of facilities and equipment for broadband	Communities with a population less than 20,000 or an urbanized area adjacent to a city greater than 50,000	25/3
Telecommunications Infrastructure Loans	1949	1936 Rural Electrification Act	$1.725 million, which allows for loans of $690 million	Construction, maintenance, improvement, and expansion of telephone service and broadband in extremely rural areas	Communities with a population of less than 5,000	n/a
Community Connect	2002	Consolidated Appropriations Act of 2004	$50 million, which allows for grants of $50 million*	Broadband deployment into rural communities where it is not yet economically viable for private sector providers to deliver service	Communities with a population less than 20,000 or an urbanized area adjacent to a city greater than 50,000	10/1
Distance Learning and Telemedicine	1990	Food, Agriculture, Conservation, and Trade Act of 1990	$82 million, which allows for grants of $82 million	Funds end-user equipment for education and health care	Communities with a population of less than 20,000	n/a
ReConnect Loan and Grant	2018	2018 Consolidated Appropriations Act	$600 million for FY 2018 plus $550 million for FY 2019 and $555 million for FY 2020**	Funds last-mile projects in the most un- and underserved communities	Communities where 90% of residents lack access to 10/1 and a population of under 20,000***	25/3

Source: Kruger (2019) and Gilroy and Kruger (2012).

* The original appropriation for 2019 was $30 million. This was increased to $50 million.

** The 2020 COVID stimulus package (CARES Act) provided an additional $100 million for the ReConnect Program (CRS 2020).

*** According to the USDA (n.d.), a proposed funded service must also be located in a rural area, which it defines as "any area that is not located in a city, town, or incorporated area that has a population of greater than 20,000 inhabitants or an urbanized area contiguous and adjacent to a city or town that has a population of greater than 50,000 inhabitants."

Indeed, the Farm Bill gives the secretary of agriculture the ability to adjust and readjust target speed thresholds as the needs of users change.[8]

The limited number of studies on the RUS broadband grants and loans have generated mixed results. Ivan Kandilov and Mitch Renkow (2010) found that the original 2000 pilot program was successful at bringing in new connections, which in turn brought about a "positive impact on employment, annual payroll, and the number of business establishments in recipient communities" (166). They qualify this observation by noting that the communities faring best were those already adjacent to urban areas. More critically, they failed to locate any "evidence that loans received as part of the current Broadband Loan Program have had a measurable positive impact on recipient communities" (165). These findings were seconded in a study of small broadband providers (under 250 subscribers) by Brian Whitacre and Phumsith Mahasuweerachai (2008), who found the existence of USDA broadband subsidies, either from the Community Connect grants or Farm Bill loans, did not attract small providers to rural areas. More recent studies see more positive results. Robert Dinterman and Renkow (2016) found that communities receiving a broadband loan did "experience modest, statistically significant increase in the number of broadband providers," and Amy Kandilov et al. (2017) found that broadband loans have a positive impact on crop sales.

What is clear from these studies and others to be presented in this chapter is that the data is mixed as to the benefits of these programs. What is required, therefore, and as Whitacre and Mahasuweerachai (2008) suggest, is a deep dive into the policies that govern these grants and loans. This is precisely what this chapter undertakes.

Critiques of the RUS

Performance: The American Recovery and Reinvestment Act
As noted in the introduction to this chapter, the 2009 ARRA represented the single largest public contribution to rural broadband deployment. It also demonstrated a major moment of legitimation for the RUS by Congress. Of the $7.2 billion allocated to broadband in ARRA, $4.7 billion was allocated to the Department of Commerce's NTIA for an infrastructure grant program called the Broadband Technology Opportunities Program, which focused on middle-mile infrastructure. The remaining $2.5 billion was given to the RUS

for broadband loans and grants, of which 75 percent was to be directed toward rural areas. This became the Broadband Initiatives Program (BIP), which focused on last-mile projects—bringing broadband directly to homes and businesses. Because it is also a loan agency in addition to a grant agency, the RUS was able to parlay its allotted $2.5 billion into a total of $3.5 billion in grants and loans. These funded 320 projects. The goal was for 7 million rural Americans to be connected, along with more than 360,000 businesses and more than 30,000 critical community institutions (Tonsager 2012).

The ARRA was a major opportunity for the RUS to assert legitimacy as a stakeholder and policy actor in rural broadband. However, the agency soon received harsh criticism. Both Congress and the RUS received complaints from telecommunications companies, government officials, and the press regarding the suitability of the RUS to distribute ARRA funds. As Tony Romm (2015) wrote in a *Politico* article that reviewed the RUS's broadband program, "RUS never found its footing in the digital age," compared with its historic sure-footedness in electricity and telephony. Of the lead-up to ARRA, Romm adds:

> Sometimes, RUS funded high-speed Internet in well-wired population centers. Sometimes, it chose not to make any loans at all. Sometimes, RUS broadband projects stumbled, or failed for want of proper management; loans went delinquent and some borrowers defaulted. Yet despite years of costly missteps that left millions of Americans stranded on the wrong side of the digital divide, a stable of friendly lawmakers swallowed their doubts about RUS and made sure the politically protected agency wasn't cut out of the historic stimulus effort.

This scathing article in *Politico* noted that in July 2015, 150 projects had not drawn the full amount of their funds (totaling $270 million). If these funds were not accessed by September 2015, they would be forfeit. Romm argued that the RUS was unprepared and ill equipped to mete out these loans and grants, which were meant to be allocated only to "shovel-ready" projects. Quick turnaround times (all awards had to be made by 2010 and all projects completed by 2013, with an absolute deadline of 2015) and a lack of staffing and resources, made it that much harder. Other critics of the broadband stimulus package worried that it would become a "cyber bridge to nowhere," expressing concern that the broadband portion of ARRA was rushed and ill considered (Herszenhorn 2009).

In both 2012 and 2014, the GAO joined in these critiques and chastised the RUS for failing to collect assessment data of recipients:

> The RUS has not shown how the approximately $3 billion in funds awarded to BIP projects have affected broadband availability. . . . Without reliable and regular

information on the results of BIP projects, it will be difficult for USDA, RUS, and policy makers to determine the impact of Recovery Act funds and BIP's progress on improving broadband availability. (GAO 2014, 22)

The RUS's responses to the GAO were, first, that because it was a loan agency it has no expertise in such capacities; second, the windfall of funds exhausted its human resources; and third, ARRA did not require "RUS to collect performance metrics from awardees" (GAO 2012, 12). In doing so, the RUS discursively distanced itself from the identity of policymaker and argued instead that it was simply a bank.

Ultimately, most projects were completed by the September 2015 deadline, and the RUS declared BIP a success in its final report to Congress. But not all was well. The RUS had to write off $325 million in ARRA funds because of terminated projects (approximately 14 percent of the total appropriation). This forced the agency to readjust its estimated number of new broadband subscribers from 847,239 to 728,733 (GAO 2014). This, of course, was a far cry from the goal of 7 million new broadband subscribers.

Eligibility: "The Hardest Thing I've Ever Tried to Do"

During the early days of the Broadband Loan Program, the RUS was accused by the USDA inspector general and the CRS of having such exacting eligibility requirements that many funds went unspent or underspent (Kruger 2018; USDA 2005). The inspector general noted in 2005 that because of inconsistent internal standards and best practices "$236.6 million in loans and grants intended to bring broadband service to rural communities was either not used as intended, not used at all, or did not provide the expected return in service" (USDA 2005, ii). The application process was deemed a barrier to entry for many potential applicants and so overly complex that in 2007 Congress ordered the RUS to change its practices. Interviewees from rural broadband providers for my research echoed these concerns:

> You know, of all the things I've tried to do in my life, the hardest thing I've ever tried to do is to navigate how to get RUS funding, how to get federal grant funding, so much so that a couple months ago, I went out to DC for a number of reasons. . . . And the purpose of my visit was simply this. "Look, you guys, your intentions are excellent with the grants for broadband, your intentions are great with loans and loan guarantees from RUS, the Department of Agriculture. But here's the deal. I cannot . . . afford to figure out how to navigate it, nor can I hire someone to figure out how to write the grant such that we would receive grant money, nor can I figure out how to navigate how to put myself in the position

to receive either grants or loan guarantees. And if you can tell me today how I do that, I'm all ears." (personal communication, 6/23/17)

This is a story that I came across repeatedly in my research: it is nigh on impossible to get a USDA loan, and, if you do, the reporting requirements are so intense that some have actually given the money back.

Perhaps the most troubling aspect of the application process is the financial commitment a company must make to guarantee the loan. More specifically, the *Rural Broadband Access Loan Application Guide* states that applicants must be able to demonstrate they hold at least 10 percent of the value of the requested loan in equity (USDA 2018b). This amount needs to be consistent at the time of loan closing as well. According to broadband consultant Doug Dawson,

> RUS loans only work if you are already a telephone company. They're almost impossible for anyone else to work. Again, I can't give you any names but a rural ISP that I represent who's not associated with a telephone company has been trying to get an RUS loan and they can't get over the hurdles. (personal communication, 4/3/18)

Telephone companies are typically those who have the necessary resources and capital to successfully secure a RUS loan. The CRS found the lack of cash on hand was the predominate reason why broadband loans were returned (Kruger 2019). This requirement, the CRS concludes, is particularly damaging to small providers, who, "critics assert, may be those entities most in need of financial assistance" (Kruger 2019, 19).

Overbuilding: The Third Rail of Broadband Policy
Since its initial forays into broadband deployment, one of the consistent industry critiques of the RUS has concerned the issue of overbuilding, which refers to a telecommunications provider building a network from scratch in a location where another provider already exists. In essence, this interloper is "building over" the architecture of the incumbent.

The concern is not the existence of overbuilding, but rather the use of federal resources to *fund* that overbuilding when that funding could be given to the incumbent to upgrade its network. This, according to detractors, undermines the policy goal of supporting unserved and underserved communities. In 2005, the inspector general of the USDA chastised the RUS for giving loans to providers in a community with an incumbent:

> We found that RUS has not maintained its focus on rural communities without preexisting service. Although the language of the law specifies that these Federal

loans and grants are for rural communities, RUS has codified and implemented a definition that cannot reliably distinguish between rural and suburban areas. Due to this ambiguous definition, the agency has issued over $103.4 million in loans to 64 communities near large cities, including $45.6 million in loans to 19 planned subdivisions near Houston, Texas. . . . The agency's current system for prioritizing underserved communities cannot, however, guarantee that communities without broadband access will be preferred to those already with access. (USDA 2005, 11)

This issue reared its head again during the development of the 2019 ReConnect Program, where overbuilding was a major theme among the 283 commenters. The major industry associations such as USTelecom, the NCTA—The Internet & Television Association (the pay television industry trade association), ITTA—The Voice of America's Broadband Providers (the now-defunct wireline broadband association), and the American Cable Association were united in their distaste for overbuilding with industry associations like the NTCA (which represents small telephone and broadband providers and cooperatives), individual commentators like Connected Nation, and GCI Communications. These comments were divided into two threads. Some, like the ITTA, point to Congress's wording of the ReConnect Program, which prohibited overbuilding:

The legislation's prohibition on overbuilding or duplication wisely recognizes that national broadband policy will be better promoted by prioritizing deployment to rural areas with insufficient access to broadband before enhancing networks or introducing new providers in areas that already enjoy sufficient access.[9] (ITTA 2018, 3)

Others, like USTelecom, appealed to altruism, noting the need to serve the most unserved areas and ignore areas already receiving service.

In focusing its Pilot on these [unserved] areas, RUS can ensure that it does not overbuild existing projects and also ensure that it brings broadband to those who are otherwise untargeted at this time. Overbuilding is inefficient because the benefits of connecting unserved are greater than those of establishing a second connection. In addition, by splitting a small potential subscriber base among additional firms, no firm may be able to obtain sufficient revenues to keep its rural network operational. (USTelecom 2018, 2)

That the final criterion of the ReConnect Program sides with the critics of overbuilding reinforces a policy path dependency. Here, the program has tight restrictions on overbuilding, including the exclusion of any area that has previously been awarded an RUS broadband loan (with the exception of BIP) and any state-funded areas. Moreover, for areas that received CAF II auction funding (see chapter 2), only those companies that won the auction can apply for the program. This effectively bars overbuilders in areas that received CAF II funding.

There are two sides to the issue of overbuilding. Funding is meant to support those communities without any service (unserved) or who cannot reach the RUS's minimum speed thresholds (underserved). As such, the argument goes that funding should be allocated to the incumbent provider to upgrade their networks. On the other hand, overbuilding stimulates competition, as it encourages new entrants to challenge incumbents (Sallet 2019b). The NRECA appealed to fairness and service in its comments to the RUS on the ReConnect Program, arguing its members needed funding to upgrade their networks:

> NRECA believes the legislative intent was to preclude RUS from inadvertently funding a new competitor to a preexisting RUS broadband borrower and thereby undermining the ability of the RUS borrower to repay the loan. . . . The Pilot Program should not bar an applicant from improving its own speed or quality of service. (O'Hara 2018)

Former FCC general counsel and Benton Institute senior fellow Jonathan Sallet agrees. In his path-charting report on America's broadband future, former FCC general counsel and Benton Institute senior fellow Jonathan Sallet argues that overbuilding is simply competition: "Overbuilding is an engineering concept; 'competition' is an economic concept that helps consumers because it shifts the focus from counting broadband networks to counting the dollars that consumers save when they have competitive choices" (Sallet 2019b, 32).

Taking note of the stakeholders here, we see those who have incumbent status like major telecommunications providers argue vociferously against overbuilding and appeal to discourses of law and universal connectivity. Those who might be deemed overbuilders—such as certain cooperatives—argue in favor of overbuilding and appeal to issues of fairness and competition. Caught in the middle are rural communities themselves, which seem not to care about policy so much as the practical issues of competition, price, and service:

> At a minimum, in order to ensure sufficient e-connectivity, a competitive environment is necessary. That requires at least two providers in the community that do not have data caps. Only offering service from a provider with these limitations greatly restricts rural prosperity. In fact, especially where there are higher levels of poverty, lower levels of education attainment and limited jobs opportunities,

having at least one provider that offers discounted service is critical. (Regional Rural Development Centers and the National Digital Education Extension Team 2018, 2)

Part of the problem of simultaneous underawarding and overbuilding is the very definition of the word "rural." So vexing was the ambiguity of this word that the 2008 Farm Bill ordered the USDA to reevaluate its many definitions (Food, Conservation, and Energy Act of 2008, §6108). The USDA's (2013) report mapped well over thirty different senses of the word used in department policy. The report ultimately concluded that a universal understanding of "rural" be adopted and recommended a standard definition of a community under 50,000 people. This mirrors the definition employed in the 2008 Farm Bill. Unfortunately, this universal definition was never adopted, and, as a result, the smorgasbord of definitions continues to pervade regulatory discourse both at the USDA and across federal and state agencies and departments.

Critiques over performance, eligibility, overbuilding, and definitions have plagued the RUS since it began funding rural broadband deployment at the turn of the millennium. As the following section explains, a reason for these myriad critiques is an identity crisis at the RUS. Is it a banker, policymaker, coordinator, or some combination? Answering this question may assuage some of its many critics, as it will provide the RUS strong direction and legitimacy in rural broadband funding and policymaking.

Narratives of Legitimacy

Banker

The obvious place to start to understand the RUS's role in rural broadband policy is to examine the language deployed by the agency itself. Recall the agency's response to the GAO's critique over the allocation of ARRA funds: that it is a bank, not a policymaker. It therefore does not possess the expertise of a regulatory body. The GAO (2012) accepted the RUS's response:

> RUS officials told us that because of RUS's traditional role as a loan administrator, it tends to focus on ensuring that the funding is disbursed, the project is built, and the agency is repaid, instead of tracking project outcome information. In addition, the Recovery Act did not require RUS to collect performance metrics from awardees. (12)

The narrative of the RUS as a bank, as opposed to a regulator (like the FCC), was also noted in interviews with former RUS officials. For example,

> RUS is the bank, and universal service [USF] provides a revenue stream to pay
> back the loan. In a sense RUS is like providing the mortgage on your house, and
> then part of your wages come from USF. . . . RUS provides the financing, but as a
> financial institution it requires ongoing revenue stream to determine the level of
> loan that is . . . financially feasible. (personal communication, 10/ 23/17)

While the RUS called itself a bank to successfully mitigate critiques of
regulatory neglect by the GAO, its role as a broadband policymaker remains
ambiguous, both legally (as we shall see below) and normatively (as found
in its own communications). For instance, in the same response to the GAO
where it called itself a bank, the RUS also categorized itself as a "policy, planning, and lending agency" (GAO, 2012, 32). This is a rhetoric oft-witnessed
in my research—the RUS claiming it is a bank while in the same breath laying
claim to the role of policymaker.

Policymaker

Like ARRA, the 2009 rural broadband plan (discussed in chapter 2) was an
opportunity for the RUS to solidify its position as a key architect of American rural broadband policy. The 2008 Farm Bill ordered "the Chairman of the
Federal Communications Commission, in coordination with the Secretary
[of Agriculture] . . . [to] submit to Congress a report describing a comprehensive rural broadband strategy" (Food, Conservation, and Energy Act of 2008,
§6112). The rural broadband policy opened for public comment on March
10, 2009. Within the scheduled two-week comment period, 225 comments
were filed, with submissions ranging from rural Americans to rural broadband providers, counties, cities, and major telecommunications and technology companies. Few commenters specifically mentioned the RUS, intimating
a lack of perceived legitimacy. Most commenters preferred to comment on
FCC policy—namely, universal service. Of those that did mention the RUS,
two themes emerged: authority and criticism.

First, a handful of commenters, such as the AFBF and Texas Statewide
Telephone Cooperative, recommended that the USDA/RUS become the
agency solely responsible for rural broadband:

> USDA has an office in almost every rural county in the nation. Local Rural Devel-
> opment, Farm Service Agency or Extension offices are staffed by members of the
> community and are ideal locations to disseminate and collect information about
> federal rural broadband deployment programs. . . . Therefore, USDA should be
> designated as the lead Federal agency for rural broadband deployment efforts

and the Secretary of Agriculture held accountable for the agency's actions. (AFBF 2009, 2)

This should come as no surprise from these actors, which have historically received funding from the RUS and whose cause the RUS champions. Nevertheless, it is significant because few commenters in this docket or others (such as 2015's Broadband Opportunity Council or 2018–2019's ReConnect Program) advocated for the RUS to take the lead in connecting rural America.

The majority of comments that discussed the RUS were critical of the agency. InLine Communications, for instance, recommended that it revise its definition of "rural," while WildBlue Communications accused it of bias against satellite internet providers. DigitalBridge Communications criticized the agency for the slow pace of awarding grants. Interviewees tended to agree with these critiques, most notably the slow pace of awarding grants and the previous GAO critique of grant allocation: "I think the USDA has a great opportunity to really take up the mantle and really take charge and support these places. But they, their programs just really fall flat" (personal communication, 1/10/18). This comment, which came from a digital divide solutions advocate and organizer, positions the RUS as a key policy stakeholder in rural broadband policy but not a fully successful one.

In summation, while Congress recognized the RUS as a policy stakeholder in rural broadband by assigning it a role in crafting the national rural broadband policy, the majority of commentators to the 2009 docket disagreed with Congress either through omission or critique. These mixed results were also present in my interviews, with respondents enthusiastic about the potential for the RUS's participation in rural broadband policy but critical of its efforts to date. The FCC, according to the majority of commentators, still held the reins of power to set rural broadband policy. As chapter 2 discussed in more detail, the entire exercise of the 2009 national rural broadband plan was ultimately for naught, as the final rural broadband policy report was written by the FCC and was made redundant by the NBP the following year.

Since the release of the 2010 NBP, policymakers have attempted to curtail the RUS's regulatory responsibilities. The 2018 Consolidated Appropriations Act and the 2018 Farm Bill, for instance, saw an unprecedent level of congressional micromanaging on the RUS. The 2018 Consolidated Appropriations Act authorized the agency to start the ReConnect Program. Rather than give the RUS power to craft the program as it saw fit, however, the act mandated that a community is eligible for the program only if 90 percent of the population fails to receive 10/1 service. This is a tremendously narrow eligibility requirement. Jonathan Adelstein, former head of RUS and FCC commissioner (now head of the WIA) suggested that this type of hamstringing is the result of lobbying from the major communication companies:

> We [at RUS] made policy decisions overseeing the distribution of funds under the Recovery Act [ARRA]. We had considerable discretion. Congress has since tried to put more restrictions on the discretion of RUS due to complaints from cable companies and others who believe RUS funding competes with private investment. Powerful incumbents lobbied Congress to focus RUS on areas where there is absolutely no service—which can make the business case difficult for RUS awardees. (personal communication, 10/23/17)

This is a consistent theme within the political economy of rural broadband—the tension between unbridled capitalism and market failure, and that between quarterly profit and universal service. Historically, the RUS has straddled these lines, providing funding for private companies in the presence of market failure. As Adelstein notes above, and as the 2019 ReConnect Program and 2018 Farm Bill attest, however, the RUS's power to draft its own policies has been curtailed by Congress, potentially to favor incumbent providers.

Coordinator

The RUS's telecommunications loans and grants are intimately connected with the FCC's CAF. Recall from chapter 2 that the USF—and through it the CAF—operates as an ongoing subsidy to support broadband deployment. It is funded by a levy on telecommunications companies. Companies pay into the fund (and pass the expense on to subscribers), and then competitive applications are made to fund rural broadband deployment. These funds are doled out over years and are meant to support the revenue stream of rural providers, whose profit margins would be minimal given the dearth of population. In contrast, the RUS operates various grant and loan programs meant specifically for capital expenditures (e.g., the installation of a fiber-optic network for a rural community). To be eligible for a loan, however, RUS applicants must demonstrate a consistent revenue stream, which may take into account USF/CAF funds. As such, many RUS applicants are dependent upon continued revenue subsidy from the USF/CAF. A full 99 percent of the RUS Telecommunications Infrastructure Loan Program borrowers receive USF funding, while 60 percent of BIP and 10 percent of Broadband Loan recipients receive such funds (Kruger 2018). Any adjustment to the USF/CAF, therefore, will heavily impact the RUS and its borrowers.

The FCC has twice restructured the USF: first in the late 1990s, when the program was established, and second in 2011, when the program switched from subsidizing telephony to subsidizing broadband. The relationship between the FCC and RUS is murky at best, but it is when the FCC has attempted to make changes to USF programs that the RUS becomes most vocal in its policy interventions.

When the FCC first crafted the USF policies in the late 1990s, the RUS was a consistent voice before the commission, lobbying for greater funding access for its constituents and greater coordination with the FCC. Here, it positioned itself as a crucial decision-maker in USF policy, writing "universal service is the core mission of the Commission and the RUS" and arguing that the FCC cannot diffuse USF responsibility to the states when jurisdiction lies with the federal government (notably, the FCC and the RUS). As it argued, "The [Telecom Act of 1996] calls for a coordinated federal and state universal service support system where state support mechanisms were intended to augment federal support mechanisms, not the other way around" (RUS 1999). The RUS went to pains to remind the FCC of its legislative authority in this domain:

> Congress charged . . . [the RUS] with implementing State Telecommunications Modernization Plans. Chief among the requirements of that legislation was that customers be able to receive 1 megabit/second data rates through their telephone lines. RUS implemented these requirements over a period of time and directed the requirements only at new plant to be constructed either in currently unserved areas or in system rebuilds. (RUS 2000b, 7)

In another 2000 filing to the FCC about a request from a New Mexico telephone company to be recognized as a rural provider (and therefore have access to the USF), the RUS (2000c) again emphasized its legislative authority, this time citing the 1936 Rural Electrification Act:

> The Administrator in making such loans shall, insofar as possible, obtain assurance that the telephone service to be furnished or improved thereby will be made available to the widest practical number of rural users. This language was in the 1949 Telephone Amendment. Advanced services capability was added to the Rural Electrification Act in 1993 and is found at 7 U.S.C. 935(d). (1)

In these early USF conversations, there is a clear attempt to lay a claim to regulatory responsibility that complements that of the FCC.

In recent years, the RUS has reduced its filings to the FCC, even though in 2014 it signed a memorandum of understanding with the Commission to augment information sharing.[10] Since 1996, the RUS has filed seventy-nine interventions with the FCC, but only twelve of those appeared after 2010, with six of those comments concerning the NBP. This suggests a retreat from the policy and regulatory arenas. As one RUS official told me, "To be honest with you, it just comes to where do we commit our resources. And you know . . . we just have too much other things going on right now" (personal communication, 8/30/17).

With the transition from the FCC's High-Cost Fund to the CAF, many of the RUS's constituents are worried about how this will impact their loans and any future capital investments in their networks. As one respondent said, "Companies are being directly impacted because they don't have the certainty of what are the regulations that are going to govern their access to the Universal Service Fund" (personal communication, 6/8/17). For its part, the RUS voiced this concern to the FCC in a filing on the NBP:

> [T]he FCC should consult with an affected government agency if changes to an existing FCC regulation could negatively impact that agency's broadband strategy. For example, in 2011 the FCC made significant changes to the federal USF. According to the [GAO], since that time, a majority of [the RUS] broadband borrowers have seen reductions in the amount of USF revenue they receive. In one example provided by GAO, 18 RUS borrowers lost an average of 31 percent of their USF support between 2011 and 2013, significantly impacting their ability to repay outstanding program loans. (RUS 2012)

Unfortunately, the RUS's comments seem to have gone unheeded as the FCC and the USAC tend to make unilateral changes to USF policy. In an in-depth interview, Adelstein was more to the point about the tension between the FCC and the RUS:

> The FCC did not always clearly discern its relationship to RUS. RUS provides attractive financing that stretches USF funding deeper into rural America—and the FCC is required by law to ensure USF is "predictable." The FCC considers itself the big dog, and will do what it's going to do, and doesn't want to be constrained. The staff was telling me, "We don't want to be trapped because of decisions RUS made, we need to change USF, and we can't be held hostage to the fact that RUS made some challenging loans even if they were based on what you expected was predictable USF funding." (personal communication, 10/23/17)

The RUS's relationship with the FCC is fluid and fraught. In the late 1990s the agency was much keener to assert expert status, whereas in the late 2000s it fell into the role of rural champion and advocate for its constituents. This shift in stakeholder status comes at a time when both government and rural broadband actors are clamoring for greater coordination between the agencies (see Gilroy and Kruger 2012). For instance, in 2015 the RUS was charged by President Barack Obama to lead the Broadband Opportunity Council (BOC) with a specific mandate to encourage coordination between all agencies involved in funding rural broadband projects. Underscoring the RUS's leadership, the NTCA (2015) wrote to the BOC that

> rather than "stepping on top of" or even competing with the efforts of RUS to manage federally-overseen resources to facilitate the construction of networks, the FCC, NTIA, and other agencies could and should coordinate with the RUS so that each can utilize its own expertise and operate effectively within its well-defined role, offering targeted solutions to market failure or regulatory barriers that limit the availability or affordability of broadband service. (5)

This need for stronger coordination between agencies and the leadership of the RUS was also echoed by one of my interviewees:

> I believe that officially there is collaboration between the two, but I'm kind of having a tough time finding it in reality. I would like to see them work together more closely because they are—I mean, a lot of companies, a lot of rural broadband companies accept monies from both plans. So, you would like them to work together better than they are, but that's just my wish list. (personal communication, 7/25/17)

The issue of poor coordination between the RUS and FCC remains unresolved. Still, the call for greater coordination has endured, present in the 2018 Farm Bill and in multiple bills in the house and senate in 2017–2018.[11] Coordination was also the rallying cry of 2017's Interagency Task Force on Agriculture and Rural Prosperity and 2019's American Broadband Initiative, which noted that "coordination [between the FCC and USDA] is critical to ensure that investments are not only complementary, but also that they maximize the impact of limited Federal funds" (Perdue and Ross 2019, 27). In fact, "coordination" is mentioned forty-five times in the American Broadband Initiative report. While this is somewhat over the top, the lack of coordination between the FCC and RUS means that communities are left behind. As we learned earlier in this chapter, many rural ISPs do not have the resources to cut through the clutter of opaque and oftentimes incongruous programs and are either forced to expend resources they do not have or go it alone. The history of the RUS is one of championing rural communities, but its inconsistent policy rhetoric since 2009 belies this tradition.

The ReConnect Program

In 2018, Congress awarded the RUS $600 million for a pilot broadband program targeting the most unserved and underserved rural communities. The 2017 Interagency Task Force on Agriculture and Rural Prosperity, chaired by Secretary of Agriculture Sonny Perdue, recommended that an e-connect fund be established with an allocation of $500 million. The Trump administration allocated $600 million for the program in 2018 and renewed the program in 2019 with a pledge of an additional $550 million. It was renewed again in 2020 with a commitment of $555 million. The 2020 CARES Act, the federal government's $2 trillion COVID stimulus package, allocated an additional $100 to the program (CRS 2020). The RUS has jurisdiction over what is now called the ReConnect Program[12] and has structured the program in categories. More specifically, $200 million is earmarked for grants, $200 million for a combination of grants and loans, and $200 million for loans. All types of business models are eligible, including traditional for-profits, co-ops, municipalities, states, and tribal communities. This program defines the rural as a city or town with a population less than 20,000. Applicants are ranked on a point system, with scoring criteria consisting of the level of population density (the least dense, the more points); the number of farms served; performance (projects proposing a network exceeding 100/100 will be favored); business presubscription (i.e., businesses need to sign up); the number of health care centers, education facilities, critical community facilities, and tribal lands served; and favorable state broadband policy.

In the program's inaugural year, 2019, the RUS received 146 ReConnect applications requesting more than $1 billion in grants, loans, and loan-grant combinations. The RUS began distributing funds to award-winners in September 2019, and in eight months meted out over $620 million to seventy ISPs in thirty-one states (USDA 2020). Even before award selection began, however, ReConnect received significant criticism, particularly around eligibility requirements. Chief among these complaints is the provision that funds can only be granted to applicants who can demonstrate that they are serving an area where 90 percent of the population receives less than 10/1. This is an exceptionally small margin of communities. Not only that, it assumes 10/1 as a baseline for broadband, when 25/3 is the country's current definition set by the FCC. As the NRECA explained in its comments to the RUS,

> The 10/1 Mbps standard is antiquated and does not constitute "sufficient access" to spur economic development (which is the heart of the problem in rural communities). In order to accomplish the goals of this program and to ensure that rural areas have access comparable to urban areas, the sufficiency standard for this Pilot Program should be raised to 25/3 Mbps . . . using a sufficiency standard of less than 25/3 will be a step backward and will effectively perpetuate a two-tiered system—one standard for urban centers and a lower one for rural America. (O'Hara 2018)

Congress (which set the 10/1 threshold) perpetuated the politics of "good enough" discussed in chapter 2. Here, the 10/1 threshold implies that *anything* above this floor is satisfactory. In contrast, many providers reminded the RUS that gigabit service is rapidly becoming the new norm. Similarly, the 10/1 standard departs from the FCC's definition of broadband as 25/3, leading to regulatory confusion (the aforementioned "many hands" problem of polycentric regulation) and inconsistent goals and norms. Such was the point made by DownEast Broadband Utility (2018):

> That is, rural households are deemed worthy of 10 Mbps/1 Mbps, but not worthy of 50 Mbps/50 Mbps—speeds that are easily attained with fiber. Additionally, why invest millions of dollars in infrastructures that provide antiquated services (DSL, cable)? "Invest Right—Invest Once." (1)

Making matters worse, the 2018 Farm Bill extended the "90 percent at 10/1" threshold to all RUS broadband grants (Agriculture Improvement Act of 2018, §6201(3)(B)(i)). If actualized as written, this will be a massive policy change for the RUS, where previously a community was eligible for RUS attention when upward of 85 percent of residents *have* 10/1. While the 90 percent at 10/1 threshold in both the ReConnect Program and the Farm Bill achieves the policy goal of serving the most unserved communities, it excludes those communities wishing to upgrade from DSL (which can meet the 10/1 threshold) to fiber or fiber-based fixed wireless. It therefore confines these communities to the broadband speeds of the mid-2000s.

The Reluctant Regulator

The RUS has been critiqued for its performance, eligibility standards, position on overbuilding, and how it defines the rural. In response, it has claimed to be a bank and therefore void of policymaking authority, a policymaker only in delineated circumstances, and a sometimes coordinator when ordered to be. These actions and critiques have led to an inchoate set of policies and regulations regarding rural broadband funding and have caused confusion among rural broadband providers and rural communities. From this confusion, however, two trends become clear. First, the RUS's perceived role in rural broadband policy shifts depending on the context and other actors in the conversation. Second, the RUS's perception of itself as a bank/lender is the most consistent description of its roles. Still, with wide responsibilities ranging from defining "broadband" and "rural" to managing a billion-dollar portfolio, its legal authority and narratives of legitimacy suggest it is much more than a bank. The RUS (2002) said as much in a comment to the FCC:

> Since the passage of the Telecommunications Act of 1996, RUS has taken an active role on behalf of rural Americans by commenting on the actions taken by the [FCC] as it has implemented the universal service provisions of the Telecom Act. Throughout this process, RUS has worked to represent the interests of all rural Americans, not just those served by RUS-financed companies and cooperatives as the financing available under the Rural Electrification Act is intended to benefit all rural areas. (1).

Moreover, as Black (2008) reminds us, banks are themselves part of the regulatory and policy process: "The 'regulators' are the banks, regulating both themselves and others to ensure compliance with the principles, at least in the initial loan documentation" (142).

We come back to the questions that define this chapter: What is the RUS's role in rural broadband policy? And how does the agency legitimize this role? This is where stakeholder analysis is put to good use. Stakeholder analysis is literally the mapping of policy stakeholders—defined as "people, groups or organizations with a vested interest in the outcome of a particular policy" (Van den Bulck and Donders 2014, 88)—along with their political and ideological positions and claims to legitimacy vis-à-vis a particular policy issue. In addition, stakeholder analysis looks at the particular views of stakeholders: their rationale, arguments, logic, and paradigmatic positions (Van den Bulck 2012).

Stakeholder analysis tends to group organizations into categories based on levels of perceived and actual interest and influence. Because assessment is based on perception, a stakeholder's place in a policy issue is dynamic—at one moment it may have high legitimacy and high power ("promoter"), while at other times it may have low power and low legitimacy ("apathetic"). It may also exist in between: high influence but low power ("defender"), or high power and low influence ("latent"). In terms of the language of stakeholder analysis, the RUS is often a "latent" stakeholder—one with the potential for high power but without consistent action (Van den Bulck 2012).

A more nuanced stakeholder typology is offered by Ronald Mitchell, Bradley Agle, and Donna Wood (1997). According to them, stakeholders fall into one of eight categories, comprised of three primary categories, four intersecting categories, and one nonstakeholder category (see figure 3.1). The primary categories include *dormant*—high power, low legitimacy, and low urgency;

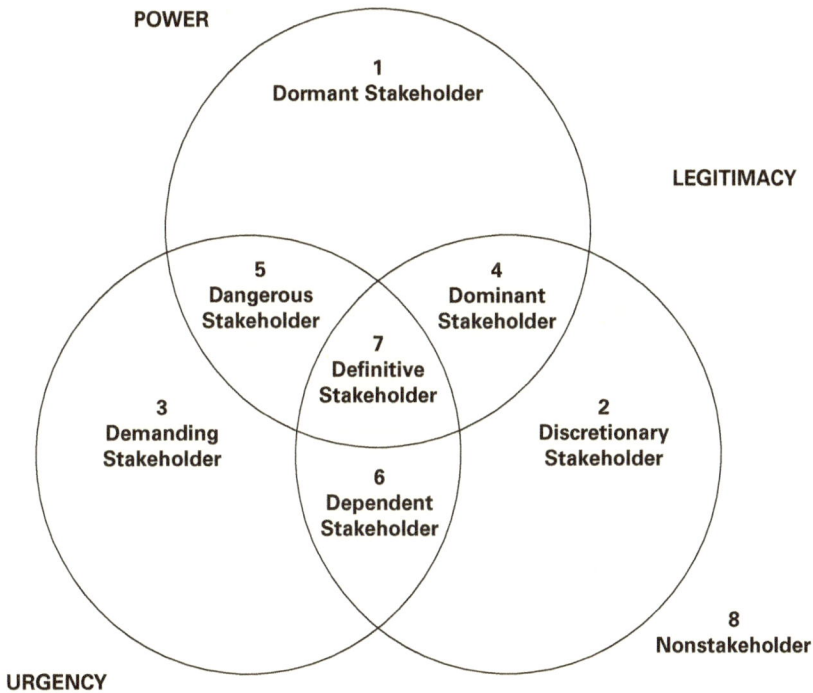

Figure 3.1
Stakeholder typology.
Source: R. Mitchell, Agle, and Wood (1997).

discretionary—low power, low urgency, and high legitimacy; and *demanding*—
low legitimacy, low power, and high urgency. When a stakeholder occupies
but one of these categories, they are considered a latent stakeholder. The
four other categories are formed by the intersection of the three primary
categories. Definitive stakeholders, for instance, stand at the intersection of
high power, legitimacy, and urgency.

The RUS's rhetoric in its FCC filings, in concert with the public comments
to federal rural broadband policy initiatives and my stakeholder interviews,
suggests that the agency fluctuates between the dormant, discretionary, and
definitive categories (see table 3.2). The RUS is definitive—high legitimacy,
urgency, and power—when Congress recognizes it through appropriations or
duties, or when an outside actor recognizes the agency as the primary mover
in rural broadband policy. At other times, such as the lack of acknowledgment by the GAO in various reports on broadband, the RUS falls back to
dormant—high power but low legitimacy—or nonstakeholder. At still other
times, such as in 2009 with the publication of the rural broadband strategy
report, the RUS has high legitimacy, tasked by Congress through the Farm
Bill to coauthor the report, and high legitimacy for some outside actors like
the AFBF, who championed its role in rural broadband policy development.
With a deadline on the report, we can even say the RUS had high urgency.

As illustrated in table 3.2, the RUS's stakeholder status in rural broadband
policy is fluid, ranging from a perceived nonstakeholder to one intimately
involved in the crafting of American broadband policy. These findings align
with Black's (2008) conclusions that stakeholder status and legitimacy are
never static but rather fluctuate in time and by actor. Depending on the
issue and the actor, the RUS's position in the stakeholder typology changes,
as does its own response to these changes.

Based on my stakeholder analysis and given the contrarian nature of the
critiques and narratives of legitimacy presented in this chapter, it is perhaps
too soon to label the RUS a *regulator* of rural broadband, although it does
have the authority to regulate broadband speeds for its borrowers. At the
very least, the agency is a major stakeholder, actor, and broad policymaker
in this field (if we distinguish "regulation" as official rulemaking from "policy" as goal setting [Freedman 2008]). I argue, therefore, that the RUS is a
reluctant regulator, one with the authority and sometimes legitimacy to act
in regulatory and policymaking settings, but hesitant to do so consistently
and constantly.

Table 3.2

Actor perceptions of the RUS's stakeholder status

Actor issue	Congress/ government	FCC	Rural orgs. (e.g., AFBF, ITTA)	Major telcos and assocs. (e.g., AT&T, CTIA, USTelecom)	Rural telcos and assocs.
ARRA*	Definitive stakeholder/ dangerous stakeholder**	n/a	Dominant	n/a	Dominant
2009 rural broadband strategy	Definitive stakeholder	Discretionary stakeholder	Dominant	Non-Stakeholder/ definitive stakeholder***	Dominant
USF 1990s	Nonstakeholder****	Discretionary stakeholder	n/a	n/a	Dominant
USF 2000s	Nonstakeholder	Dormant stakeholder	n/a	n/a	Dominant
2015 Broadband Opportunity Council	Definitive stakeholder	n/a	Dominant	Nonstakeholder/ definitive stakeholder	Dominant
2018 ReConnect	Definitive stakeholder	Dormant	Dominant	Definitive	Dominant

* Romm's (2015) criticism of the RUS's handling of its stimulus funding puts the agency in the demanding or dangerous stakeholder categories.

** While Congress considered the RUS a definitive stakeholder by allocating it a portion of the ARRA, the GAO (as noted above) questioned both its accountability and power to oversee the project, hence the dual categorization of definitive and dangerous.

*** Some did not mention the RUS in their comments at all, suggesting the agency is not a priority. Others, such as the NCTA, mentioned the RUS to critique the opacity of the funding programs.

**** Derived from GAO reports on universal broadband, which omitted the RUS entirely.

There are important implications of this reluctancy, the most pressing of which is delays to broadband deployment in rural America. The RUS has faced constant criticism for its lack of transparency and accountability; mismanagement of funds; inability to agree on key terms, definitions, and eligibility criteria; the pace of funding; and its bureaucratic opacity. Nevertheless, Congress and borrowers trust the RUS to shepherd the country through the digital divide. The discrepancy between perception, practice, and policy needs to be resolved at the legislative level by Congress and at the internal level by the RUS itself to avoid past mistakes and improve efficiency. Failing to clarify its own perceptions risks the RUS repeating past mistakes and contributing to an opaque regulatory and funding structure that hinders broadband deployment. It furthermore risks exacerbating what Jonathan Koppell (2005) calls the "multiple accountabilities disorder" of polycentric regulation, wherein regulators battle for legitimacy and power in an unclear regulatory landscape.

As much as I have described regulatory and policy discord, there is also vast potential for policy alignment. The RUS has both the perceived and actual legitimacy to take a leadership role in rural broadband deployment. This position was echoed during in-depth interviews when I asked if the RUS is able to craft a national rural broadband plan:

> On the question of is RUS the best agency to lead this, it is. It certainly is. Rural Development and the USDA programs, they know rural communities best.[13] The companies, the cooperatives that are accessing these loans, they know rural development. There's a presence in almost every state across the country. People are comfortable with these individuals, and that plays such an important role in really working through what often become very complex application processes. (E. Frederick, personal communication, 1/18/18)

The RUS has a legislative mandate to champion rural America and has offices and divisions in every state. It understands the needs of rural communities better than any federal department or agency. Charging it with the responsibility to steer America's future rural broadband plans would also recall an earlier time in our nation's history when Congress trusted the USDA to connect our rural communities with electricity and telephony through the REA.[14] In this book's conclusion, I revisit this history and the findings in this chapter to argue that the RUS and USDA should author and direct America's rural broadband plan for the 2020s.

Conclusion: Of Groats and Bits

Rural broadband is a lot like the grain grown and harvested throughout the American Midwest. Grains—such as wheat, oats, rice, and corn—are considered food staples throughout the world. Some grains, like oats, are part of an essential diet, while others, like barley for beer, can be considered guilty pleasures. Cooperatives were early ways farmers organized to bring their harvests to market (Keillor 2000). The same applies to rural broadband. The original infrastructure was laid by rural residents in the form of co-ops, which have become central to its distribution. Broadband is used for the necessities of rural life (education, business transactions, telemedicine) and for guilty pleasures (Netflix bingeing, porn). Both grain and broadband also share the unfortunate domination of large corporations (Crawford 2019; Hendrickson, Howard, and Constance 2017).

Groats (the kernels of cereal grains) and broadband bits also share the presence of the USDA and RUS. The difference in this capacity is that with oats, cereals, and agriculture more broadly, the USDA and RUS understand their roles. With broadband, their roles are more ambiguous, which is reflected in their policies and processes. Because of this ambiguity, the FCC has taken the lead in rural broadband deployment policy. It does so in a way that privileges large, incumbent telecommunications providers over local ISPs and cooperatives historically championed by the RUS.

Eighty years ago, the RUS helped bring electricity and telephony to America's hinterlands, making sure those removed from the centers of power, commerce, and education were not left behind. This leadership in rural communications has waned in the last two decades, however. Today, with control over a billion dollars in loans and grants and a trusted presence throughout the country, the RUS plays an important, if ambiguous, role in the struggle to connect millions of rural Americans to the digital ecosystem. Understanding, clarifying, and improving the role of the RUS and thereby recovering this lost leadership potential is crucial if the country is to realize the goal of universal broadband.

4 "It's Pretty Cool": Cooperatives, Municipalities, and Rural Connectivity

The Peoples Rural Telephone Cooperative (PRTC) in McKee, Kentucky, might have the best name of all broadband providers in the country. The name captures the collective determination that saw rural America empower itself through telecommunications in the New Deal. PRTC was formed in the 1950s, and six decades later embraced the opportunity to leverage federal funds to bring fiber to an area that some have started to call "Silicon Holler" (Rosenblum 2017).[1] With ARRA funding (see chapter 3) and a team of mules to pull the wires, PRTC now offers gigabit connections to every home and business in Jackson and Owsley Counties—some 18,000 customers. These are two of the most economically deprived counties in the state (Miller 2019). As PRTC's CEO Keith Gabbard is fond of remarking,

> We don't have a four-lane highway, no hospital, no college, no railroad. We don't have any of those things in these two counties but we have as good a broadband as they've got in any city in the country. It's pretty cool. (personal communication, 2/5/18)

The story of PRTC reminds us that, at its core, rural broadband is not about policy but about people. It's about the quality of life we enjoy, the information we consume, the businesses we run, and the health care we receive. It speaks both to the future for an impoverished community and the future of connectivity. Indeed, it reminds us that while federal policy is crucial, rural broadband is *lived* by those people and companies serving rural communities in geographically embedded places. As we learned in chapter 2, the large national providers disfavor rural broadband deployment and have been less responsive to community needs than consumers desire. Instead, rural cooperatives, municipalities, and local ISPs are leading the charge for rural connectivity. All of this is embodied in the PRTC.

Yes, broadband is about people—users, businesses, and providers—but the story of PRTC also reminds us about something else: that broadband is fundamentally local. By this, I mean the technologies of access, be they mobile, fiber, or wireless, terminate at a geographically local destination. The companies successfully closing the digital divide are locally based, be they cooperatives, municipalities, or local providers. State-level policies, when articulated correctly, empower local communities and cooperatives. Local partnerships, organizations, and community members become champions of connectivity; and, of course, the users are local, for we are always in situ, in place, even when we are saving our pages to the cloud (Casey 2013).

That "all broadband is local," as Olivier Sylvain (2012) powerfully argues, epitomizes my argument that the solution to the rural-urban digital divide will be found not in the boardrooms of national providers and global telecommunications firms but at the local level. Throughout rural America, communities have come together to address their own digital infrastructure needs. These communities have gone about it themselves by developing municipal broadband plans, mobilizing electric and telephone cooperatives to take on the burden, and organizing public-private partnerships to roll out fiber or fiber-fueled fixed wireless networks. Many rural communities are navigating the inchoate array of federal policies, programs, and subsidies described in chapters 2 and 3 and are partnering with state rural broadband offices, and with each other, to meet their constituents' digital needs. In doing so, they are the unsung heroes of rural broadband.

The academic literature on cooperatives often invokes the idea of a "cooperative commonwealth"—a social-democratic alternative to the machinations of neoliberal capitalism that privileges quarterly profits over long-term vision and community solidarity (Keillor 2000; Schneider 2018). With rural broadband we see the emergence of a *connected cooperative commonwealth*, either in actual electric and telephone cooperatives or simply in the form of neighborly cooperation. As Eric Ogle (2019), senior consultant at Magellan Advisors, recently pondered, "What if rural communities begin thinking about broadband competition as a way to keep money circulating through a local economy?" (38). Rural communities are not just *thinking* about these microeconomic initiatives, they are *living* them through broadband deployment.

In 2019, there were over 560 communities with some form of municipal network—meaning networks that have been at least partially funded by local public dollars (more on this below). These include:

- 63 municipal networks with publicly owned FTTH
- 63 communities with a publicly owned cable network
- 237 communities with some publicly owned fiber
- more than 120 communities with publicly owned dark fiber
- more than 230 communities in thirty-three states with a publicly owned network offering at least 1 gigabit services (Community Networks 2020)[2]

In addition, most telephone cooperatives and over 100 electric cooperatives are now offering retail broadband services (out of 260 telephone and 834 electric cooperatives, respectively) (Trostle et al. 2020).[3] Like their municipal network counterparts, these cooperatives often offer fiber and gigabit connectivity and meet or exceed the speeds offered by the incumbent (which is usually a cable company or a telecommunications company offering DSL) for a fraction of the cost. These companies keep connectivity local, meaning local presence, local service, and, most importantly, local accountability.

This chapter chronicles the efforts of rural communities and cooperatives to connect the unconnected through a case study of one rural, agriculture-based community in southwestern Minnesota. Through an ambitious initiative that bonded the future of the county's wind energy surplus to fiber, the county partnered with Alliance Communications, a South Dakota-based telephone cooperative, to connect the county with a fiber-optic network. Today, the entire county (with the exception of the county seat of Luverne, for interesting policy reasons) is served or has the potential to be served with FTTH. This rural fiber-optic network has attracted interest from major business—a key aspect of rural development for the "Nutcracker Capital" of the Midwest.

The example of Rock County underscores the two fundamental arguments of this chapter: that broadband is local and that cooperatives are key stakeholders in America's digital ecosystem. It furthermore exemplifies what Elizabeth Roberts et al. (2017) call "resilient" rural places, "whereby local resource are developed so that rural communities have the capacity to steer wider processes in a global context and highlighting the non-linearity, processual and messiness of rural places" (373). ICT, most notably broadband, are vital to this practice of resiliency (Ashmore, Farrington, and Skerratt 2017; Roberts et al. 2017) and to rural development in general (Klein 2016; Malecki 2003; Malecki and Moriset 2008). Connecting this chapter to the larger argument of my book, the success of future projects like

Rock County requires the democratization of the entire broadband policy system, particularly around funding programs and state prohibitions on municipal networks.

I begin with a discussion of "broadband localism" (Sylvain 2012) and challenge the seemingly inextinguishable belief that the internet represents a "global village." Instead, the internet, and broadband connectivity more specifically, represents a very local phenomenon, of which co-ops are crucial stakeholders. These are member-owned companies that began as electric or telephone cooperatives funded by the REA in the 1930s, 1940s, and 1950s. Their very names recall these earlier endeavors: the Peoples Rural Telephone Cooperative in Kentucky, Farmers Telecommunications Cooperative in Alabama, and Citizens Telephone Company in Georgia.[4] Today, hundreds of co-ops provide high-speed broadband to parts of the country neglected by the major national incumbents. They demonstrate how the solution to rural broadband deployment lies not exclusively, or even primarily, with the national providers but at the local and state levels, empowered by coordinated policies at the federal level. I capture these dynamics through a case study of Rock County, Minnesota, and the Rock County Alliance, and ultimately propose what I call the "rural broadband model," which emphasizes the importance of spatiality, technology, usage, structure, and governance to successful rural broadband projects. For clarity, I use "local" here in the simplistic and geographic sense of the term—referring to a territorially bounded place—rather than broader interpretations of it.[5] I conclude with a discussion of states' roles in empowering local communities and how a strong federal policy apparatus is necessary to make any of this happen.

Broadband Localism

We are still in the clutches of the half-century-long metaphor of Marshall McLuhan's (1964) global village—a unified world that communicates in the clouds and through the ether. While poetic, the internet is anything but ethereal. Instead, the materiality of the internet—namely, our broadband connections—defies the argument that advanced telecommunications has left us with a condition of "no sense of place" (Cairncross 1997; Meyrowitz 1985). If anything, broadband reminds us how place-based we actually are.

Delving deeper, we often think of advanced capitalism—of which ICTs are a key component—of epitomizing Karl Marx's ([1939] 1993) "annihilation of space by time" or David Harvey's (1991) more contemporary "time-space compression." To a certain extent, this is correct. For those select privileged few almost every part of the world is accessible—if not physically, then certainly digitally (Castells 2010). But it is also true that the architecture of the internet and the deployment of broadband has given us the world at our fingertips in a fundamentally local way. Indeed, the architecture of the internet is such that broadband access—our gateway to the internet—is a fundamentally place-based and local phenomenon (Hu 2016; Starosielski 2015). My argument that broadband is local is not unique. Other scholars, noted below, have made this argument. What is unique is the broad scope of my argument, as others who have invoked the idea tend to explore it narrowly, focusing individually on technology, spatiality, usage, structure, or governance. As I demonstrate in this chapter, all of these elements factor into broadband localism.

Technology: The Last Mile

Broadband is local because the technological points of access are local. We are always *somewhere* when we access the internet, no matter if that connection is wired or wireless. Wired connections, for instance, must terminate in a home, business, or anchor institution:

> All broadband is local. The speed and bandwidth capacity of local network infrastructure determines users' media consumption habits. They affect whether users socialize through email, social networking, or video chat. They define whether users turn to cable television or their laptop to watch video programming. (Sylvain 2012, 796)

Tony Grubesic and Elizabeth Mack (2017) describe how the termination spots for wireless connectivity are, like wired broadband, fundamentally local: "Simply put, the bulk of wireless systems remained hardwired, at least somewhere in their network architecture, but their wireless access points remain highly localized" (42). Local termination is known as "the last mile"—the connection between the consumer and the ISP (Grubesic and Mack 2017). The cable coming into your home or business—be it DSL, coax, or fiber—is the last mile even if it's rarely a mile in distance. The signal from the tower to our phone is also the last mile. The last mile refers not to distance and not to technology, but to a relationship between the ISP and the end user—a relationship that is by definition local.

Spatiality: Broadband in Place

Grubesic's (2003, 2006, 2008, 2010) work also demonstrates how decisions about broadband deployment are local in that they are tied to an intersection of markets and space. Grubesic (2008) has spent the past twenty years mapping broadband deployment, concluding that while deployment is typically concentrated in major metro markets, it has spillover effects on neighboring and sometimes rural communities. There exist broadband cores and peripheries, with notable "islands of availability," where a community has broadband but surrounding areas do not, and "islands of inequity," where a community lacks broadband but is surrounded by areas where broadband has been deployed. While the latter can often be newly constructed suburbs and exurbs, the former demonstrate that "certain local markets have been successful in attracting infrastructure investment and broadband upgrades" (Grubesic 2006, 443).

Grubesic's (2006) work on cores and peripheries builds on earlier work on what he calls "broadband regions." These are not isolated to metro areas. Instead, he notes that communities with more competition in broadband (broadband cores) will experience greater technological advancement than those with less competition (broadband peripheries). While perhaps obvious, these findings violate the teleological argument that advanced technologies develop first in metro areas, then move to suburbs, exurbs, rural regions, and finally remote regions. While islands of availability exist throughout the country, including rural areas, greater competition in urban, suburban, and exurban communities (and, of course, the wealthy areas therein) might mean that rural and remote regions fall so far behind that they never catch up. As Grubesic (2006) concludes, "It is becoming increasingly clear that the broadband periphery, those areas with limited levels of competition, are beginning to display a certain amount of geographic inertia [meaning resistance to change]" (445). To Grubesic, a lack of competition and a lack of regulatory attempts to stimulate competition may be more predictive of a long-lasting digital divide than efforts to fill the gaps. Jan van Dijk (2020) agrees, noting that despite advances in broadband deployment, the digital divide may never be closed because of the growing global levels of inequality.

The more "remote" a community is, the less competition there is, and the less it benefits from what we might call markets of proximity, meaning less broadband availability (Townsend et al. 2013). This recalls Kyle Nicholas's (2003) phrase "geo-policy barriers," whereby geography acts as a form of capital, either permitting or prohibiting telecommunications deployment. Geo-policy barriers "are chokepoints, mechanisms of control created through the interaction of geography, market forces, and public policies" (Nicholas 2003, 287). In a study of telephone access in Texas, Nicholas found that while public policy worked to correct the rural penalty, a negative externality developed in the form of a "remote penalty," leading remote communities to self-connect through a local co-op rather than wait for an incumbent to connect them.

In sum, geography and morphology—collectively "spatiality"—are critical to an understanding of broadband localism. As Grubesic (2010) argues, "Broadband provision is an inherently local problem" (130). Solving this local problem requires local solutions.

Usage: Adoption and Digital Inclusion

In the introduction, I reviewed how the presence of broadband benefits rural communities. There, I introduced my notion of the five pillars of rural broadband: economic development, education, telehealth, civic engagement, and quality of life. These pillars speak to the usage and adoption of broadband within local communities, so it is unnecessary to review them here. All of the research and policy on deployment is for naught if people do not actually *use* (or "adopt" in policy parlance) broadband.

Structure and Governance: Municipalities and Policy

Sylvain (2012) coined the term "broadband localism" to capture the shift from popular thinking about the internet as an amorphous and metaphysical cloud to companies and policymakers realizing that "residency matters" in terms of connectivity. Sylvain's use of the term is confined to the experience of municipalities in broadband deployment, most notably as network builders and providers. This conceptualization thus shifts the conversation away from technologies of access, markets, and usage to organization, structure, governance, and policy. Municipal broadband, as noted in the introduction to this book, is defined by the presence of a local government in the funding, planning, construction, and/or operation of a broadband network. In Europe, this is often known as "community broadband" and has been notably studied in the United Kingdom (Ashmore, Farrington, and Skerratt 2017) and elsewhere. Many have pointed to municipalities as being the lynchpin in correcting the failures of the private market (Crawford 2019). As Susan

Crawford wrote in her 2019 book, "It turns out that America's awful, expensive data connectivity is a national problem for which the solution is intensively local: cities and localities are leading the way" (67). To this, Sylvain (2012) adds,

> Local governments are lighting the spark for broadband infrastructure build-out. They are mobilizing an array of local anchor institutions and resources to bring service to residents. That they do this is no surprise. After all, local governments are best suited to appreciate the characteristics or "terroir" that distinguish their constituents from others. (805)

According to these scholars, not only are communities solving the digital divide themselves, but they are contributing to a sense of resilience among communities and community members, especially in rural areas (Ashmore, Farrington, and Skerratt 2017).

Unlike in the United Kingdom and the EU, in the United States substantial disagreement exists as to the legality of municipalities offering broadband services. At least nineteen states have enacted legislation either prohibiting municipalities from funding and operating broadband networks or creating inhibitive barriers, such as massive feasibility studies and provisions against using public funds (Baller Stokes & Lide 2019).[6] The issue for many regulatory scholars is the FCC's power to preempt these prohibitive state laws under its authority to ensure universal service (via §706 and 254 of the Telecommunications Act of 1996) (J. Cobb 2018; Dunne 2007). In an earlier article, I noted this tension between local, state, and federal policy and jurisdiction, concluding that the FCC's inaction on municipal broadband was foreshadowed in earlier decisions not to intervene to help local municipalities (Ali 2017b). Matthew Dunne (2007) goes further, arguing the FCC has a duty to advocate on behalf of municipalities:

> When localities, presumably best positioned to appraise local needs and services, attempt to accelerate broadband deployment and investment in information infrastructure, the Commission is empowered under section 706 and Title I of the 1996 Telecommunications Act to protect the localities in their efforts. (1163)

Unfortunately, the FCC's major attempt to preempt state barriers to municipal broadband (in Tennessee and North Carolina) failed when the Sixth Circuit Court vacated the FCC's decision in 2016.[7] Since then, the FCC has not rekindled the issue. Some commissioners have been downright hostile to the very thought of municipalities becoming providers. Take FCC commissioner Michael O'Rielly, for example, who called municipal broadband a threat to the First Amendment: "In addition to creating competitive distortions and misdirecting scarce resources that should go to bringing broadband to the truly unserved areas, municipal broadband networks have engaged in significant First Amendment mischief" (qtd. in Bode 2018a). Municipal broadband, of course, is just the opposite of a threat to the First Amendment, as it gives those without connectivity the ability to enter the digital public sphere.[8]

There is considerable debate in the research and in scholarly communities about the feasibility and sustainability of municipally funded broadband networks. Proponents, such as the Institute for Local Self Reliance (ILSR) and Next Century Cities (NCC), champion the rights of municipalities to connect themselves and offer dozens of case studies and success stories (see, e.g., Muninetworks.org). Legal scholars similarly argue for the right of municipalities to offer broadband (Dunne 2007), and others argue that municipal broadband corrects market failures (Crawford 2019; Grubesic and Mack 2017). Interview-based studies demonstrate the qualitative importance of municipal broadband for "community resilience" (Ashmore, Farrington, and Skerratt 2017). Critical political economists argue for stronger public options generally, seeing public investment as the only way to correct the digital divide (Pickard and Berman 2019).

In contrast, critics see municipal broadband projects as risky financial endeavors with little benefit and the potential for great harm (e.g., Yoo and Pfenninger 2017). Many of these conclusions are derived from white papers and research reports published by institutions including university law schools such as New York University (Davidson and Santorelli 2014) and the University of Pennsylvania (Yoo and Pfenninger 2017), conservative think tanks (e.g., State Government Leadership Foundation; see Ford 2016), and think tanks with strong ties to the telecommunications industry (e.g., Phoenix Center; see Ford and Seals 2019). Ford and Seals (2019) found that, despite calls to the contrary, municipal broadband (in the form of a case study of Chattanooga, Tennessee) does not increase labor market participation. In a 2016 study, Ford questioned the "economic advantages resulting from government-owned broadband," arguing that municipal broadband does not increase competition and therefore "is a very inefficient way to obtain the positive externalities of broadband" (8). Ford (2016) goes so far as to argue that "the economic development from municipal broadband systems is based on stealing business from other cities" because businesses are more likely to relocate to a municipality with a faster network rather than start up because of a municipal broadband system (17). He concludes that municipal networks should only be contemplated "where private entry is not profitable." Charles Davidson and Michael Santorelli (2014) agree, finding that "the direct economic impact of GONs [government-owned networks], especially in job creation, can be difficult to attribute. Data do not indicate that GONs actually serve as the nucleus of renewed economic activity in cities and towns where they have been deployed" (xiv). While endorsing public-private partnerships as a viable solution, they conclude that many municipal networks are borne from the "mission creep" of local utilities and that "GONs are not remedies for perceived or actual broadband connectivity challenges" (103). The same conclusion is reached by Steven Landgraf (2020), who argues that the entry of a municipal broadband network actually serves as a disincentive to private network investment.[9]

Yoo and Pfenninger's (2017a) now infamous study has become the focal point in the research debate over municipal networks. The authors analyzed the public financial disclosures of twenty municipally funded FTTH projects, concluding that the projects, and others like them, are unlikely to succeed financially. They estimate that it will take some municipalities over a century to repay loans and debts. As a result, Yoo and Pfenninger (2017a) caution city leaders to "carefully assess all of these costs and risks before permitting a municipal fiber program to go forward" (23). Taxpayers will bear the burden of the risk, with defaults likely leading to higher taxes or lower quality services.

Yoo and Pfenninger's study generated a heated debate over its intent, method, rigor, and conclusions. Both the ILSR (C. Mitchell 2017) and New America (Null and Nasr 2017) published scathing rebuttals to the report, focusing particularly on data collection and method. In return, Yoo and Pfenninger published both a correction to their initial report (2017b) and a rebuttal of their own (2017c).

These debates, which are often ideological in nature, come down to the question of whether broadband is a consumer good or an essential utility and basic infrastructure like the highway system. Critical political economy, with its focus on the democratic allocation of vital resources, would find agreement in the latter. As Crawford (2019) poignantly argues, "Ultimately, last-mile fiber is a public good. It will not pay for itself quickly—we never expected that of the highway system—and should not be expected to" (212). Christopher Mitchell (2017), director of the Community Broadband Networks Initiative at the ISLR, agrees, arguing that precisely because there is controversy in this space, municipalities need to make the decision regarding broadband themselves, without state or federal interruption. He suggests that if "local leaders have taken the time to gather all the relevant facts and can evaluate all the pros and cons of an investment before making any commitments," then they should be free to make the decision that's right for their community. To this point, Roberto Gallardo (2016) reminds us that two hallmarks of rural America—decentralization and localism—are embedded within broadband architecture. This demands a local solution to a fundamentally local problem. And this is exactly the approach taken by Rock County, Minnesota.

Rock County, Minnesota

Rock County sits in the southwesternmost corner of Minnesota, bordering South Dakota on the west and Iowa on the south (figure 4.1). It is home to 9,490 people, 4,619 of whom live in the county seat of Luverne. Spanning 483 square miles, the county has a mixed economy, with health care and social assistance, retail, manufacturing, and agriculture being the most common industries. Its population is 94.1 percent white, 2.82 percent Hispanic or Latino/a, and 1.01 percent Black or African American. The county's median household income is $56,753 against a statewide median of $68,388. Its poverty rate is 11 percent, 1 percent higher than the statewide level of Minnesota. While Rock County boasts the most valuable farmland in Minnesota, its name is derived from Rock River and the prominence of quartzite in the region.

Luverne was one of four towns profiled in Ken Burns's sweeping World War II documentary *The War* and refers to itself as the Midwest's "Nutcracker Capital," thanks to 2,500 nutcrackers donated to the Rock County Historical Society by its president, Betty Mann. Rock County is also famous for another reason: it is the most connected county in the state, with 99.93 percent covered by a fiber-optic network managed by Alliance Communications, a telephone cooperative based out of nearby Garretson, South Dakota.

The story of Rock County and what came to be known as the Rock County Alliance both embodies and engenders my broadband localism thesis. It speaks to the importance of community champions, partnerships, municipalities, cooperatives, states, and fiber. These are all seminal qualities

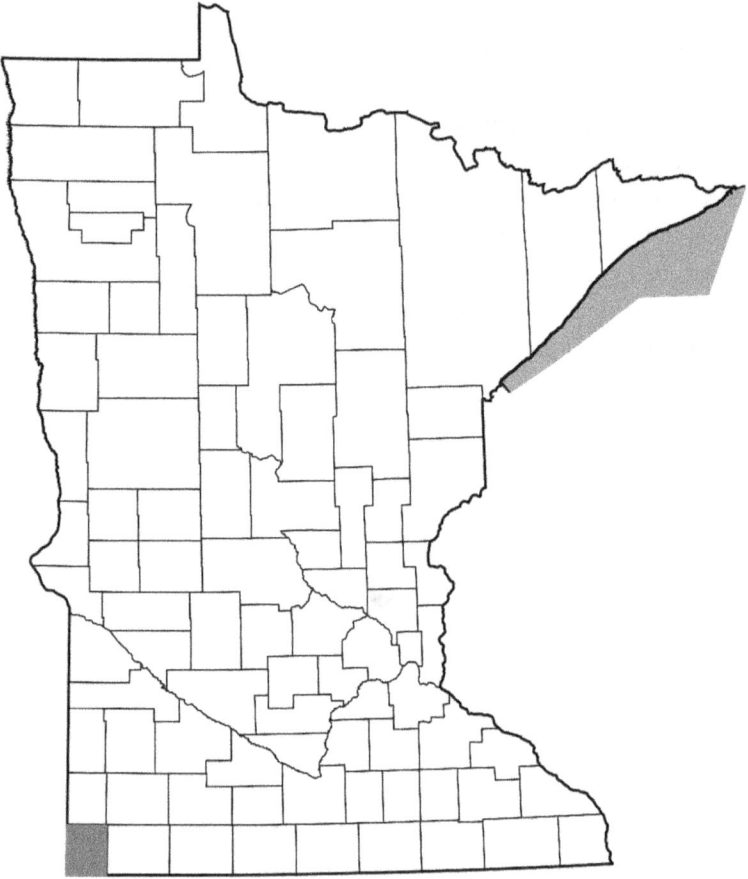

Figure 4.1
Rock County, Minnesota.
Source: *Wikipedia* (2020).

necessary to connect the unconnected in rural America. Importantly, this
is a story of not only success but also challenges, reminding us that the path
to connectivity is fraught with legal impasses, financial risk, and numerous
false starts. Still, it is the story of only one rural county in America. As was
often said to me during my interviews, "If you've been to one rural community, you've been to one rural community," or "You talk to one cooperative,
you've talked to one cooperative." As critical rural studies teaches us, collaps-
ing rural narratives into a singular, homogenous experience does nothing to
foster rural voices (Thomas et al. 2013). That Rock County embodies several

facets of rural connectivity, however, makes it is useful to highlight the dynamics of broadband localism. To avoid essentializing, however, I will at times draw from my other interviews and field visits to bolster my arguments and expand these narratives.

Early Steps

Rock County's first foray into high-speed connectivity came in 2009, when the Woodstock Telephone Company (now Woodstock Communications), based out of neighboring Pipestone County, won an ARRA award to bring fiber to fifteen communities in three counties (Pipestone, Rock, and Lyon). The award totaled just over $15 million, including a $10.6 million grant and a $4.6 million loan from the USDA's BIP (see chapter 3). While Rock County would certainly benefit, it was not involved in the planning or application. Unfortunately, the project did not materialize, as costs skyrocketed by millions in unexpected fiber and labor expenses. As recalled by Kyle Oldre, county administrator for Rock County and one of the key informants for my research, the 1931 Davis-Bacon Act, or prevailing wage rule, needed to be fulfilled, since federal funds were involved. The Davis-Bacon Act requires that laborers working on public projects using public money for projects over $2,000 need to be paid according to the local prevailing wage, determined by the state's department of labor. These wages are typically higher than those in more rural areas, leading to accusations that prevailing wage laws increase the cost of federally funded building projects (CRS 2012). There is substantial disagreement, however, as to whether prevailing wage laws actually raise the cost of projects or instead lead to a more qualified workforce, greater productivity on the job, and greater spending power of workers (Mark Price and Herzenberg 2011). In a comprehensive review of scholarship on the topic, Kevin Duncan and Russell Ormiston (2019) conclude, "Prevailing wage laws do not affect construction costs, promote worker safety and training and do not have a racially discriminatory impact" (153). Regardless of the debate and cause, however, Woodstock saw an increase in project costs by several million dollars, for which it was unprepared.[10] Woodstock asked the three counties for help in meeting these costs, but all three refused. In 2012, two years after being issued the grant/loan combination, Woodstock had yet to draw on the loan, making this "one of the few BIP projects nationally" that had not done so, according to the Governor's Task Force on Broadband (GTFB 2012, 41).

Kyle Oldre has seemingly endless energy and enthusiasm for his county. Working out of the county courthouse—a red brick mansion in the center of Luverne—he is credited by his peers as being the person who brought fiber to Rock County. Broadband champions are an essential first step in a community's decision about broadband (Gallardo 2016; NCC 2018). "The biggest lesson for us, as a foundation, is you have to have a local champion," said Bernadine Joselyn, CEO of the Blandin Foundation, a Minnesota-based nonprofit dedicated to improving the life of rural Minnesotans (personal communication, 1/15/19). Roberto Gallardo, in his book *Responsive Countryside* agrees wholeheartedly: "it is critical to have at least one trusted local champion. . . . *If the will and motivation exist, ways to get things done will be found*" (2016, emphasis added). Roberts et al. (2017) and Fiona Ashmore, John Farrington, and Sarah Skerratt (2017) call these individuals "digital champions." For Rock County, Oldre is that champion and after the disappointment of Woodstock, he made it his mission to find a new dance partner.

Finding this partner was a challenge and emblematic of the market failure of rural broadband. As Oldre told me when asked about the primary challenge for Rock County,

> Finding a partner. Really, it is finding a partner. It's really trying to convince somebody that you're worth investing in. You might believe it, and I might believe it, that Rock County is an absolute great place to build a fiber network, but until you can get those people that actually own those networks to believe that same thing, there's no way to do it. Because we met—prior to this, we met with every dance partner we could think of. (personal communication, 7/24/18)

The largest companies were unwilling to service the county, and although Luverne was served by two cable companies—Vast and Mediacom—their networks did not extend past city limits. As Oldre recalled, Rock County's efforts were quickly dismissed by the major telecommunications and local cable companies:

> They were like, "Are you kidding me?" So they turned us down. And then we met with the CenturyLinks and we met with Verizon to see if there was something they could do better to enhance us. They were like, "No." We were too small. We didn't matter in the scope of things. And that's hard to say about the community you care about. (personal communication, 7/24/18)

Fortunately, Rock County suffered only from industry *neglect*. Other municipalities across the country seeking to deploy broadband were not so lucky. More often than not, when governments—be they municipal or county—contemplate broadband provision, their efforts are met with intense industry *backlash*. Often this is done with the argument that local government intervention in broadband provision "distorts" the free market (Ali 2017b). Of course, the argument is for naught, considering that the industry—be it telecom or cable—was failing to provide service, or adequate service, in these communities in the first place. Nevertheless, telecommunications companies have been unceasing in their lobbying of local governments and state legislatures against municipal broadband projects. In Fort Collins, Colorado, providers spent over $900,000 trying to block a proposed city-owned network (Chamberlain 2019). While they failed in Fort Collins, they have proven victorious on numerous other fronts. So far, nineteen states have enacted laws prohibiting or inhibiting municipal broadband (Baller Stokes & Lide 2019). This comes amid a total of $92 million in lobbying at the national and state levels by the telecommunications industry in 2018 (Chamberlain 2019). State laws range from outright prohibitions in Texas and Nebraska to inhibitive barriers in terms of feasibility studies in Virginia to confinement to city boundaries in North Carolina.

Minnesota has some barriers to overcome for municipal broadband, including the need for a referendum and a supermajority of 65 percent (Chamberlain 2019). Still, it is not as bad as other states. In Minnesota, the bigger issue for municipalities considering broadband is industry hostility. Renville and Sibley Counties, some 140 miles northeast of Luverne, for instance, received pushback when they launched a hybrid public-private fiber network called RS Fiber. An op-ed by Brent Christensen, president and CEO of the Minnesota Telecommunications Association, exemplifies this pushback:

> From the very beginning, the RS Fiber project was an ill-conceived idea that left taxpayers in 10 cities and 17 townships at considerable financial risk if the network failed to meet its business plan. Today, there is plenty of evidence that project proponents ignored warnings that cost estimates and revenue projections were overly optimistic. Heck, one proponent even stated that the network would be profitable by 2018! (qtd. in Treacy 2018b)[11]

It is ironic that industry, and industry-supporting FCC commissioners such as Michael O'Rielly quoted earlier, push back so aggressively on municipal broadband projects yet simultaneously fail to deliver adequate service to these broadband deserts. As we learned in chapter 2, the FCC is content to dole out subsidies to the major telecommunications companies while failing to take into account constructive alternatives like municipalities and cooperatives in stemming the tide of the digital divide. Luckily, Rock County was spared this vitriol.

Searching for a Partner

After the Woodstock flirtation, a nibble came from neighboring Lismore Cooperative Telephone Company, some twenty miles from Luverne. It wanted to deploy a fixed wireless system in Rock County, based on a fiber ring connected to a series of towers. Lismore had been successful in deploying fixed wireless systems in neighboring Nobles County and even started laying fiber to select areas. But Rock County wanted fiber to the home and wouldn't settle for anything less. This illustrates a key tenet of municipal broadband: that municipalities need to assess their own communication and infrastructure needs (Gallardo 2016; NCC 2018).

Another challenge for Oldre was convincing his county board of supervisors of the need for high-speed connectivity:

> I had some commissioners that didn't really think we needed internet. He didn't have a cell phone and . . . he just didn't see the value in it, so he wasn't going to support any investment in it at all. And it's actually when I had a board turnover—guys like Jody—and now, it's like, OK, we need to do some things out there and really step up to the plate and be involved and not just sit back and expect the private sector to deliver it to us. When the reality is we need to do something ourselves. (personal communication, 7/24/18)

Oldre is referring to county commissioner Jody Reisch, who quickly became a digital champion for Rock County after he moved out of Luverne and realized he couldn't get internet access: "Doesn't everybody have it?!" he joked to me. Outside of Luverne, the fact was that very few in the county were connected. Oldre calls Reisch the county's first digital champion:[12]

> And so, he buys this beautiful house just outside of town—no internet. And he's like, "I can't believe it." He assumed that there would be internet out there. Buys this beautiful house [*laughter*], now he can't do his job. So he's looking at offices and he says, "We've got to fix this problem." So he was really the champion. And now, of course, he's got speed like you dream about. (personal communication, 7/24/18)

Together, Oldre, Reisch, and commissioner Stan Williamson started pounding the pavement, trying to find a dance partner. Finally, in 2012 or 2013 (the year is unclear), after a regional development meeting and another round of failed conversations, the trio decided to drive to South Dakota and cold-call Alliance Communications. As Oldre recalls,

We get in the truck and we're like, "Dammit." And Stan says, "Well, I know this guy in Garretson. We can go talk to Don" [Don Snyder, then general manager of Alliance Communications in South Dakota]. So really we cold-called him. We drove from Slayton, through Rock County, over to Garretson, middle of the afternoon and said, "Hey, Don. What do you think about building fiber for us?" And that's when he sat back in his chair. "Yeah, right. Five million bucks" [*laughter*]. (personal communication, 7/24/18)

Oldre, Reisch, and Williams all hinted this $5 million was a total guess on Snyder's part. As Reisch tells it: "And I think Alliance, when they made the initial offer, they had an idea of what it was going to cost. I think they threw out a BS number [*laughter*]. I don't think they anticipated us hitting the goal" (personal communication, 7/24/18).

At last, Rock County had a dance partner in Alliance Communications, a telephone cooperative located in nearby Garretson, South Dakota. Now they just needed $5 million to get started.

Minnesota is a national leader in state encouragement of rural broadband deployment (Pew Charitable Trusts 2020). In 2008 the state began investigating the need for broadband and in 2009 parlayed funds from the ARRA State Broadband Initiative to map broadband and plan for deployment. In 2011, spearheaded by the Blandin Foundation and then-governor Mark Dayton, the state organized the GTFB. This came with an ambitious set of goals, including a 2015 goal of universal 10–20 Mbps download/5–10 Mbps upload (GTFB 2012). Having achieved this goal (90.77 percent coverage in 2018), the state's current goal is for all residents to have 100/20 speeds by 2026 (GTFB 2018). This is all part of Minnesota's Border-to-Border broadband plan and the self-proclaimed Minnesota Broadband Model, which includes

- realistic, forward-looking internet speed goals
- an Office of Broadband Development
- broadband deployment mapping capabilities to accurately plan, monitor, and track broadband infrastructure
- the Border-to-Border Broadband Development Grant Program. (GTFB 2018, 7)

Back in 2012, a key recommendation of the task force was the creation of a broadband office. This materialized in 2013 with the creation of the Office of Broadband Development, under the auspices of the Department of Employment and Economic Development. Danna Mackenzie was hired to run the operation. The state also allocated $20 million for broadband grants starting in 2014, with the stipulation that no award surpassed $5 million and that grants could cover up to half of eligible infrastructure costs. Areas were eligible if they fell under Minnesota's definition of unserved, meaning without a wireline service of 4/1. Rock County (with the exception of the city of Luverne) met these requirements hands down and was awarded a $5 million grant for broadband deployment—the largest grant awarded in the 2014 cycle and exactly the amount they needed.

Cooperatives

Cooperatives are incredible organizations: member-owned and community-driven, they are fundamentally local and are credited with giving rural communities a voice in an uncompromising and urban-normative capitalist system (Curl 2010; Keillor 2000; Schneider 2018). Cooperatives respond to market failure, opening up markets in rural areas where the erstwhile free market dared not tread. In the history of American agriculture, and in the Midwest especially, cooperatives were cornerstone institutions, bringing rural areas retail stores, grain elevators, creameries, telephones, and insurance when the logic of the free market denied such services (Keillor 2000). Minnesota even became known as the "cooperative commonwealth" for the number and diversity of its cooperatives. As the Farmer–Labor Party of Minnesota wrote in 1935, "The Cooperative Commonwealth is Minnesota's American solution to the American problem brought on by a predatory, ruthless capitalism" (Day 1935, qtd. in Keillor 2000, 312).

Most cooperatives are based on the Rochdale system, a set of seven rules that trace their origins to the 1844 weavers of Rochdale, England:

1. Voluntary and open membership
2. Democratic member control (one person, one vote)
3. Member economic participation
4. Autonomy and independence
5. Education, training and information
6. Cooperation among cooperatives
7. Concern for community (International Cooperative Alliance, n.d.)

Adding to this, Steven Keillor (2000) sees a three-pronged definition of cooperatives. There's an economic definition, "whereby members by-pass the market adjacent" to them. There's also a democratic definition, wherein a cooperative "is also a polity governing its own affairs." Last, there's the social definition, in that cooperatives are "characterized by mutual expectations" (Keillor 2000, 6). In other words, in economic, democratic, and social terms, cooperatives are inherently, distinctly, and proudly local.

As discussed in chapter 1, cooperatives were essential to the wiring of rural America with both electricity and telephone (Kline 2000; MacDougall 2014). These initiatives were supported by a robust and committed federal policy system developed and nurtured by the REA. It should come as no surprise, then, that when the free market abandoned them, Rock County turned to a cooperative "to create a marketplace in rural areas that had none" (Keillor 2000, 7).

For rural, sparsely populated communities, a partnership with a cooperative is ideal. Technology is expensive and expertise tough to come by. But a co-op, be it electric or telephone, has the infrastructure and organizational abilities. In 2020, some 109 electric cooperatives and most of the 226 telephone cooperatives offer retail broadband, and these numbers are growing (Trostle et al. 2020). As Oldre said to me about Alliance Communications, "They're a rural co-op and this is kind of their specialty. They know that they can make it work, and they're not worried about 20 cents a share return or anything like that. It's staying within the co-op" (personal communication, 7/24/18).

Two issues of accountability are present here: local and fiscal. In the first regard, cooperatives have become key to the deployment of rural broadband because they are, by definition, local. As Oldre noted to the question of whether he had faith in Alliance,

> Oh yeah. But part of it was, we knew they were in Garretson. I wasn't signing a deal with some guy in Texas. I was signing a deal with a neighbor. There's board members that have to still live in the community if they screw this. So yeah, there was faith, but it was not a crazy amount of faith. It was if they told us they were going to do something, we had every reason to believe them. (personal communication, 7/24/18)

Oldre also mentioned another key aspect about cooperatives: they understand that the return on investment for major infrastructure development needs to be measured in decades, not years. One of the seminal critiques against major telecommunications company is their short-term, shareholder-driven need for return on investments. The larger telecommunications companies are unable (or unwilling, if we're being cynical) to invest in the "long game" of broadband. Simply put, shareholder-driven companies are driven for one reason: returning dividends back to shareholders. To this point, Century-Link recently announced that it is pulling back in rural communities and refocusing on so-called enterprise arrangements (Dampier 2018). Similarly, AT&T has largely abandoned DSL and will concentrate on cellular and fixed wireless networks for rural areas (Brodkin 2020b; Engebretson 2016). As recounted in chapter 2, CenturyLink also failed to meet its CAF II buildout requirements in Minnesota in 2018, while Frontier failed in Nebraska and New Mexico (Taglang 2019).

Co-ops, in contrast, understand that their first duty is to the community and its members. This translates not into immediate financial returns but rather into services rendered. This means that co-ops are able to invest in areas where market failures exist because they are not expecting a one- or two-year return on investment but rather a five-, ten-, or even twenty-year return. As Danna Mackenzie exclaimed, "I think co-ops are just such a wonderful model for rural areas. They have more patient capital, they're owner-investors, they're supporting the service that is in service of themselves and their neighbors" (personal communication, 6/29/18).

This long-term vision also makes them risk-adverse—a double-edged sword when it comes to entering a new market like broadband. As Nathan Schneider (2018) explains in his book *Everything for Everyone,*

> When all a business does is serve its existing members, and those members' perceptions of their worlds, it can fall stagnant, and stagnation in a changing world is no help to members. Investor-owned businesses attract investor-owners on the premise that they'll forever exceed expectations; it is both a virtue and a peril that co-ops tend to find their members under the humbler premise of meeting known, day-to-day needs. (233)

The adherence to the status quo reminds us that the cooperative broadband movement is not all sunshine and roses. Recall from the introduction to this book that there are two types of co-ops involved in broadband: electric and telephone (there is also RS Fiber, a "broadband-only" cooperative, but it is an outlier). While both are intimately connected to the REA, they could not be more different. On the one hand, telephone cooperatives have been providing telecommunications services to rural areas for a century. They are experts in both deployment and customer service. To paraphrase what one respondent said to me, you don't call the electric company when your toaster is broken, but you do call the telephone company when your internet is down. Telephone cooperatives are also exceptionally well organized nationally through the NTCA—The Rural Broadband Association. On the other hand, telephone cooperatives are used to relying on both capital and operational subsidies (from the RUS and FCC, respectively), which may breed fiscal caution. In Christopher Mitchell's opinion, electric cooperatives have been more aggressive and are better placed to bring broadband to rural America:

> There's a turf war now. If you're a rural telephone company or a rural telephone cooperative and you're facing an electric utility offering internet service, you should probably be afraid. As a big fan of co-ops, I don't like to say negative things about them, but the rural telephone co-ops are nowhere near as efficient as the rural electric co-ops in my estimation because . . . the rural electric cooperatives have never received operating subsidies. They've received some capital subsidies, but they always had to make their business models work. (personal communication, 11/20/17)

Adding to this is the need for "smart grids"—electrical infrastructure that is networked via fiber optic connections (Tucker, Goodenbery, and Loving 2018).[13] This means that they have miles of fiber already at their disposal, connecting hubs in electricity distribution and sometimes community anchor institutions. Mitchell continues,

> Where there's an electric coop offering the same services as a telephone coop, I think the electric coop has an advantage. That's led to these fights in USDA . . . not just [over] the telecom grants and loans . . . but also the smart grid, because the telephone companies and the telephone cooperatives really don't like the federal government giving a lot of smart-grid money out to the electric co-ops because they can use that to build networks that they will also offer telecom services on. (personal communication, 11/20/17)

While Mitchell sees the future of rural connectivity residing with the fortunes of electric cooperatives, they, like their telephone counterparts are also risk averse. Their national organization, the NRECA, has been reticent about fully supporting retail broadband. As a senior official said to me,

> We have a broadband resolution that urges us to advocate for advanced communications systems for our members both . . . on the electric utilities side, the backbone, and then, to the extent that our members want to go into the broadband business, provide them resources and partnerships, etc. to do that. We don't necessarily encourage it, but if they want to go into the retail business, we help them find solutions. (personal communication, 8/22/18)

He continued, indicating there may be a shift in the wind:

> We are encouraging our members to look at advanced communications as a piece
> of their needed electricity infrastructure investment as you go into sort of the smart
> grid of the future and what that means, of smart meters and command response,
> etc. And to the extent they want to leverage the investment they make on the
> backbone side, the electric utility side, and then help their consumer members get
> access to broadband. We're really encouraging that. But their business is electricity
> and it has historically been electricity. (personal communication, 8/22/18)

A reason for this hesitancy is the ambiguity of state law. As noted earlier,
several states have laws that prohibit electric cooperatives from providing
services beyond their mandate (i.e., electricity), while other states have left
the area ambiguous. As the ILSR (2019) concludes, "Co-ops in many states
are loathe [sic] to challenge state law in court" and as a result many of these
laws exist despite shaky legal footing. Fortunately, numerous states, including North Carolina, Georgia, Mississippi, and Missouri, have loosened these
restrictions, giving electric cooperatives the much-needed breathing room
to contemplate expansion. In fact, when Missouri established its broadband
office in summer of 2018, one of the first steps was to pass House Bill 1880
to encourage rural electric cooperatives to provide fiber-fueled broadband.

Despite their differences, electric and telephone cooperatives have been
enormously successful in deployment broadband to rural America. Rock
County was thus in good hands with their dance partner, Alliance Communications, in 2013.

Implementation of the Fiber Network

Alliance Communications has gone by a number of names in its 100-year
history. Its current iteration came into existence in 2003, when Baltic Telecom and Splitrock Telecom merged. Today, the Garretson-based company
serves the southwest pocket of South Dakota, including Sioux Falls, along
with areas in southwest Minnesota and northwest Iowa. Alliance was therefore not unfamiliar to the residents of Rock County, as the towns of Hills
and Steen were both served by the co-op. The extension into Rock County
was a natural extension to the fiber Alliance was laying in South Dakota.
Minnesota's Border-to-Border grant program was what sealed the deal.

With the $5 million grant secured from Minnesota's broadband office,
work between Rock County and Alliance Communications began in earnest in 2013. Because Minnesota law forbids grants from going to co-ops not

located in the state (private LLCs are not affected by this law), Rock County and Alliance created a new privately held, Minnesota-based company: the Rock County Broadband Alliance (d/b/a Alliance Communications) in 2014.

Rock County insisted the entire county be served with fiber, rather than just the densest areas, and it estimated this would cost $12.8 million. Of that, $5 million came from the state grant, while the rest would come from Alliance. Yet again, however, the prevailing wage issue reemerged, as Minnesota has its own state version for state-funded projects. According to Kyle Oldre, cost estimates increased by over $2 million. The partners discussed cleaving off two of the twelve townships slated to be on the network to reduce costs, but the county board knew that if they were omitted, these townships would never get service, As Oldre explained, "Population density's so small, they'd be flyover land forever" (personal communication, 7/24/18).

To rectify this financial shortfall, the county embarked on a risky million-dollar bond. Fortunately, as Oldre and Reisch recall, if they had to put that much money down, the timing couldn't have been better because the county was just starting to receive payments from its wind production tax. This tax nets the county about $600,000 a year. The county bonded this tax to $1 million ($100,000 a year for ten years) to come up with the $1 million necessary to pay the prevailing wage increase to the broadband budget. While $1 million may seem like a drop in the bucket to the digital divide (to which $80 billion would be needed to solve [de Sa 2017]), in a county that only collects $5 million from taxpayers annually, it is a tremendous sum. More than the amount, however, it was crucial this money not come from taxpayers, nor would it come only from Luverne (the largest city in the county). As Oldre recounted,

> We're not going to have to go out there to taxpayers and raise property taxes. It was a great investment, because this money's coming from the rural segment. It's not like I'm taking Luverne's city taxes and applying it to something outside Luverne. So I'm not raising the taxes for the folks that might not see the direct benefit. It's funny: it's the only project I've ever been part of where we spent a million dollars and I have zero complaints. We've had nothing but compliments. (personal communication, 7/24/18)

Rock County is not unique in its decision to bond tax dollars to finance fiber deployment (Crawford 2019). Renville and Sibley Counties and the eight towns that make up RS Fiber did the same thing. The difference is that the partnering cities were required to back the bond and were "asked to agree to raise taxes to make the loan payments if the cooperative does not make them on their behalf" (Busch 2018). This means that if the broadband provider—RS Fiber, a unique, broadband-only cooperative—were to default, taxpayers would be on the hook for a decade (Busch 2018).

Unlike RS Fiber, Rock County does not own its network. This was done to adhere to both Minnesota and county law. Minnesota is a Dillon's Rule state, meaning that the county can only do what is specified in law (rather than do anything that is not specified or prohibited).[14] Accordingly, they had to justify the financial outlay as a public purpose. Funding the network certainly constituted such a purpose. Operating and owning the network, however, were deemed beyond the scope of public purpose. As a result, "The county released any ownership in the fiber the minute we wrote the check for a million" (Oldre, personal communication, 7/24/18). Their faith was entirely placed in Alliance Communications.

Construction on the Rock County Alliance fiber network began in April 2016 with the ambitious goal of complete deployment in one year. Construction began in the northern part of the county, known as Jasper, and fanned out from there (see figure 4.2). Because of a stipulation in the state grant that the money could only be used for areas considered "un- or underserved," the county seat of Luverne was left off the plans, since Luverne was served by two cable companies. This left Luverne, for the time being, as what Grubesic (2006) calls an "island of inequity"—a community with poor service surrounded by communities with excellent service.

Over the course of 2015 and into 2016, Rock County Alliance ramped up a publicity campaign to address the dual challenges of its quick buildout and the lack of visibility of Alliance throughout the county. The campaign and its motto, "Wired Differently," proved remarkably successful. As Ross Petrick, CEO and general manager of Alliance explained, unlike in many communities where a publicity campaign must double as an education campaign to teach communities about fiber and broadband, Rock County residents already knew the value and potential of high-speed connectivity: "The need was just—it was there. They definitely understood the need for broadband. And some communities, back when they started this, they didn't understand the need for broadband, but Rock County did" (personal communication, 5/1/19). Perhaps this awareness contributed to the fact that when Alliance began deployment, the take rate (the percentage of residents promising to subscribe) was just shy of 90 percent. This is an

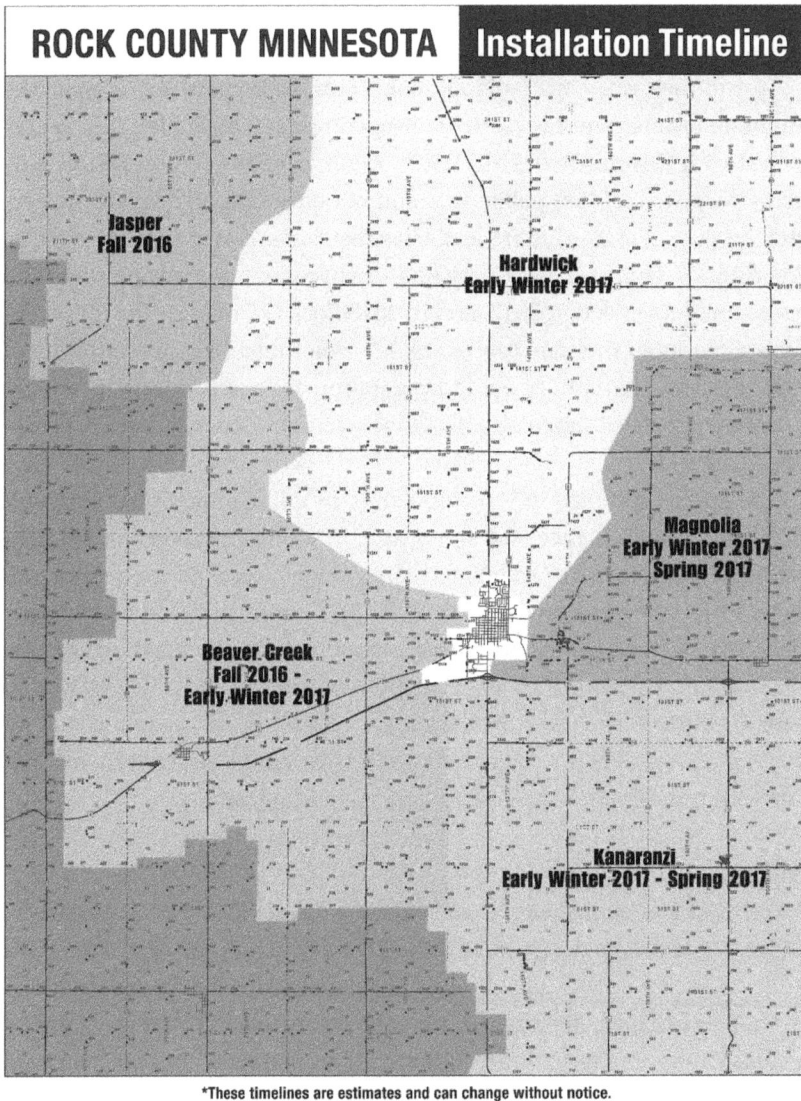

*These timelines are estimates and can change without notice.

Figure 4.2
Installation timeline for Rock County Alliance broadband network.
Source: Kyle Oldre and Ross Petrick.

incredible accomplishment and one that far surpassed the expected rate of 60 percent of eligible customers.

True to their word, the buildout took a year, an impressive feat given the Minnesota winter and the quartzite peppering the topography. Two factors helped with this timeline. First, Rock County gave the project countywide right-of-way to build on public land and not have to secure easements. Second, because of the mapping and assessments performed back in 2009 by Woodstock Telephone, the county was effectively shovel-ready for fiber. As a result, buildout ended in 2017, and in 2018 Rock County was the most connected county in the state (see figure 4.3), boasting a 100/20 availability of 99.93 percent. This is an amazing accomplishment and connectivity rate for a densely populated urban area, let alone a county with only twenty people per square mile!

The project completion is memorialized by a literal rock, engraved with the name of the Alliance and now sitting in the county office building in Luverne (see figure 4.4). Overall, 531 miles of fiber were laid and service provided to 1,100 customers at a cost of $23,652 per mile and $11,417.55 per customer. The average monthly subscription is currently $100, which gets you 100 Mbps download. The entire project cost came to $12,559,307.92, with an estimated return on investment of 9.51 years—a time frame no doubt inconceivable to a major national telecommunications company but one within the comfort zone of Alliance Communications.

Impact

The impact was immediate. In one example, KQAD-FM, on the outskirts of Luverne, had been paying upward of $2,000 a month to their provider. With Alliance, their monthly bill is around $85 per month (*MSBA Advocate* 2017). In another instance, doctors are now able to take fuller advantage of digital records. "I was unable to do any clinic work from home," wrote Dr. Richard Morgan of the Sanford Luverne Clinic to the Minnesota government. After Alliance Communications came knocking, everything changed: "Our new reliable broadband service means that I am able to access my patients' electronic medical records from my home at any time of day" (Morgan, n.d., 1).

The wackiest boon came in November 2018, when Tru Shrimp announced that Luverne would be the site for their new 12,000-square-foot "harbor" (shrimp incubator). Numerous studies have demonstrated that the presence of advanced telecommunications facilities is a primary driver in a

2018 Broadband Availability in the State of Minnesota

Percentage of Households Served by Wireline
Broadband Service by County
At Least 100 Mbps Download/20 Mbps Upload Speeds
Statewide Availability: 74.11%, Rural: 60.05%

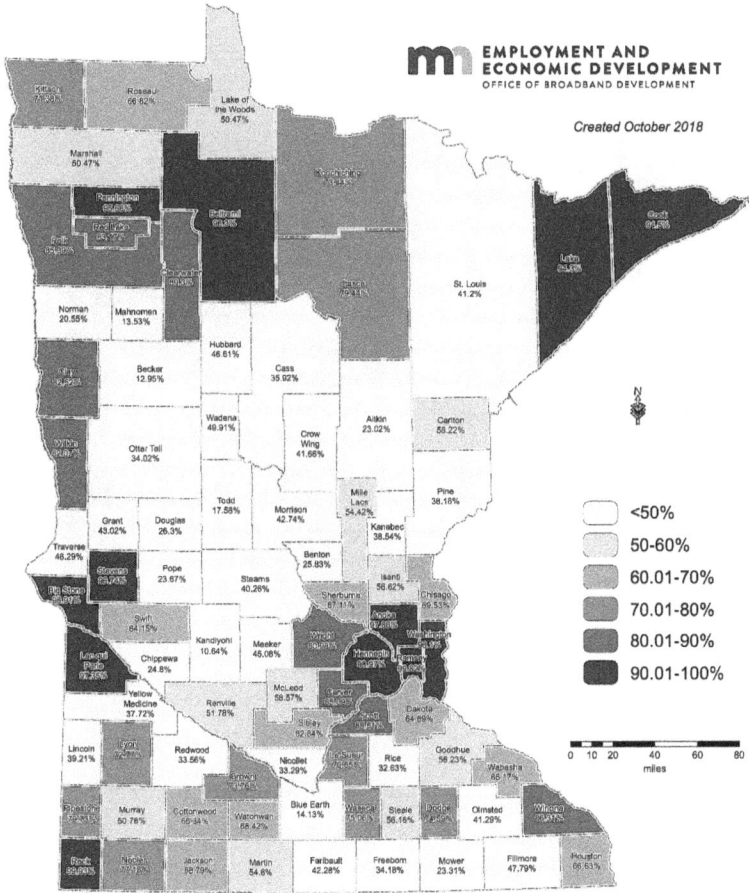

Figure 4.3

Broadband availability in Minnesota at 100 Mbps/20 Mbps, 2018.

Source: Minnesota Employment and Economic Development (2019).

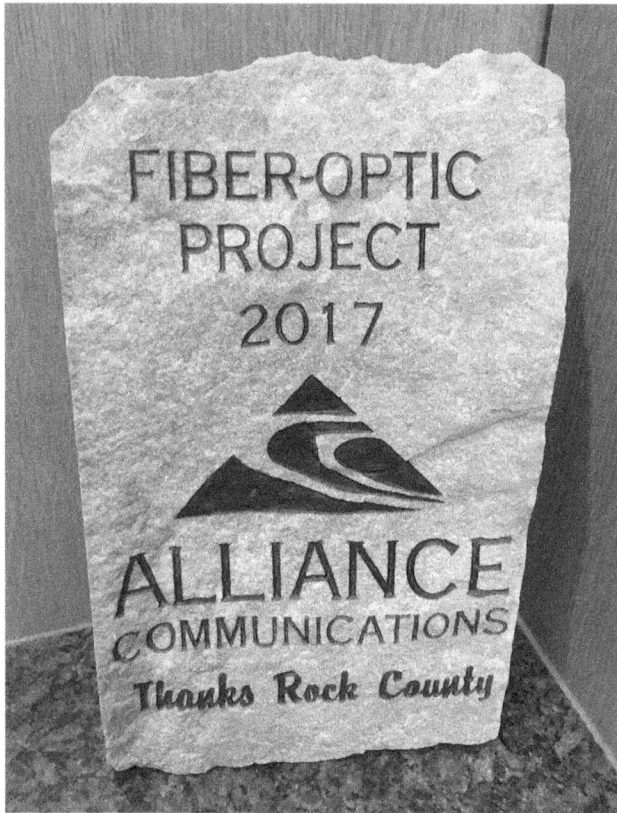

Figure 4.4
The Rock County Alliance rock.
Source: Author.

firm's decision to relocate (see, e.g., Lawless and Gore 1999; Townsend et al. 2013), and this was no exception when it came to Tru Shrimp. The facility, whose groundbreaking would occur in 2019, promised to bring sixty jobs and $27 million in economic stimulus to the county. While this partnership never materialized (not because of anything Rock County did, but because of certain environmental standards in Minnesota), it demonstrates the importance of fiber connectivity for the future of rural business.

In 2018, Rock County won a Broadband Communities grant of $75,000 from the Blandin Foundation to work on digital inclusion and skill development. The Blandin Foundation is one of the leading proponents of digital inclusion in Minnesota and has worked with both the state and local governments to improve digital connectivity, access, and adoption throughout the state. The Blandin Broadband Communities program is a two-year partnership between rural communities and the foundation to "define [communities'] technology goals, measure current levels of broadband access and use, and seek technical assistance and resources." Blandin has helped communities get Wi-Fi on school buses and helped local businesses "claim their space on Google Maps" and enhance their websites. "Every community's unique," said Bernadine Joselyn, Blandin's director of public policy and engagement. What they have in common is the need for broadband: "It's just an absolute necessity to have access to the internet and the skills to use it. But without that, communities are just going to fail in their attempts to retain, let alone attract, knowledge workers, young people, employers, employees" (personal communication, 1/15/19). Rock County's Broadband Community grant went to the local library—a lovely building located in the heart of Luverne, run by Calla Jarvie.

Not to be left out, Alliance is now expanding its fiber network into Luverne, financing this expansion itself. Since Alliance is technically an "overbuilder" in Luverne, it is not eligible for grants or loans. According to Ross Petrick, this is all part of Alliance's mission to "build in"—closing the gaps in service, ensuring that all communities are served, and working on digital inclusion plans to make sure that customers are getting the most out of their connections.

Minnesota

Kyle Oldre credits much of Rock County's success with timing. The Minnesota grant, the partnership with Alliance, the community take-up—everything came together when it needed to:

> If I was going to do this again in 2018, if I had it—I couldn't pull it off. Because I wouldn't—the competition has gotten ruthless on where the broadband dollars are going. It's become much more politicized. It used to be, "Hey, you know we're going to [be] geographically diverse, we're going to spread around." We didn't worry so much about the numbers. As a county of 10,000, as a county of 30,000. It was just about trying to do good things. I'm not sure that we would be as competitive today as we were then. (personal communication, 7/24/18)

Oldre is underselling the important role that he, Jody Reisch, and Stan Williamson—Rock County's digital champions—played in connecting this sparsely populated county. We also cannot discount the role that Minnesota—the country's leader in state rural broadband deployment and policy—played.

The local problem of rural broadband in Rock County required a local solution, but it also could not have happened without the state. States are crucial stakeholders in the push for rural broadband (Pew Charitable Trusts 2020; Whitacre and Gallardo 2020). This occurs both legislatively and discursively. The mere act of state representatives talking about and promoting rural broadband is an important first step. When it comes to legislation, Minnesota demonstrates how five dimensions are necessary to create a robust local broadband ecosystem: an office of (rural) broadband, a state broadband plan, grants, favorable municipal broadband legislation, and cooperative broadband legislation.

As noted above, Minnesota created its Office of Broadband Development in 2013. The office works with both providers and communities and, perhaps most importantly, articulates the state's broadband plan and administers the Border-to-Border grant program (Minnesota Department of Employment and Economic Development 2020) As regards Minnesota's state plan, recall chapter 2 where I discussed the need for policymakers to be technologically neutral but not technologically blind. This philosophy is practiced and preached in Minnesota, where the state does not advocate a particular technology but does advocate a particular speed threshold:

> Minnesota policy is technologically neutral. The law says we accept all technologies and that's absolutely the case. . . . The way Minnesota law is written, we can fund anything that is scalable to 100×100 or better . . . and that requires that they have to deliver 100×100 service out of the gate. (D. Mackenzie, personal communication, 6/29/18)

After having met their 2015 goal of statewide (or border-to-border) 10/5, the 2016 broadband task force set a new goal of border-to-border speeds of 25/3 by 2022 and 100/20 by 2026.

Next is funding. Between 2013 and 2017 the state awarded $85.2 million in broadband grants (GTFB 2018). To achieve the 2026 goal, and after considerable political discord, in 2019 the Minnesota legislature awarded an additional $40 million over two years for subsequent broadband grants. Last, the state has favorable legislation toward both municipal and cooperative-operated broadband networks. While Chapter 237, Section 237.19, of the Minnesota statute requires municipalities to obtain a supermajority of 65 percent of voters to initiate a broadband service, it does not stand in the way once that supermajority is achieved so long as the network does not compete with an existing private provider and "such services are not and will not be available through private telecom companies in the foreseeable future" (Chamberlain 2019; see also Baller, Stokes & Lide 2019; Minnesota Statute §429.021(19)(ii) 2020).

The case of Rock County Alliance demonstrates the importance of local communities and cooperatives to the future of rural connectivity. They are the unsung heroes in a capitalist market filled with failure and rural neglect. Minnesota permits municipalities to fund networks and cooperatives to operate them. More than that, it celebrates these digital alternatives, something that dozens of other states should follow. In my travels across the Midwest, I frequently heard other states (most notably Missouri) wanting to emulate the Minnesota model—from policy development to grants to recognizing the local and multistakeholder nature of rural broadband. Judging from Rock County's experience, it's not hard to understand why this "cooperative commonwealth" is the envy of other states.

Conclusion: The Rural Broadband Model

Each rural community is unique, and there is no cookie-cutter solution to the digital divide. Rural broadband is a local problem that requires local solutions. Reflecting on the story of Rock County and the literature on rural broadband, however, five criteria were essential to its successful rural broadband deployment. Extrapolating from this case study, I suggest that when contemplating broadband deployment, each rural broadband project needs to consider (see figure 4.5):

- spatiality—a predetermined area of deployment
- technology—preferably fiber
- usage—business and digital inclusion
- structure—cooperatives and municipalities specifically
- governance—local, state, and federal policy

These dimensions come together to form what I call the "rural broadband model" and underscore the argument that all rural broadband is fundamentally local. These local solutions, however, cannot be accomplished without robust state and federal policy mechanisms, which also factor into the model.

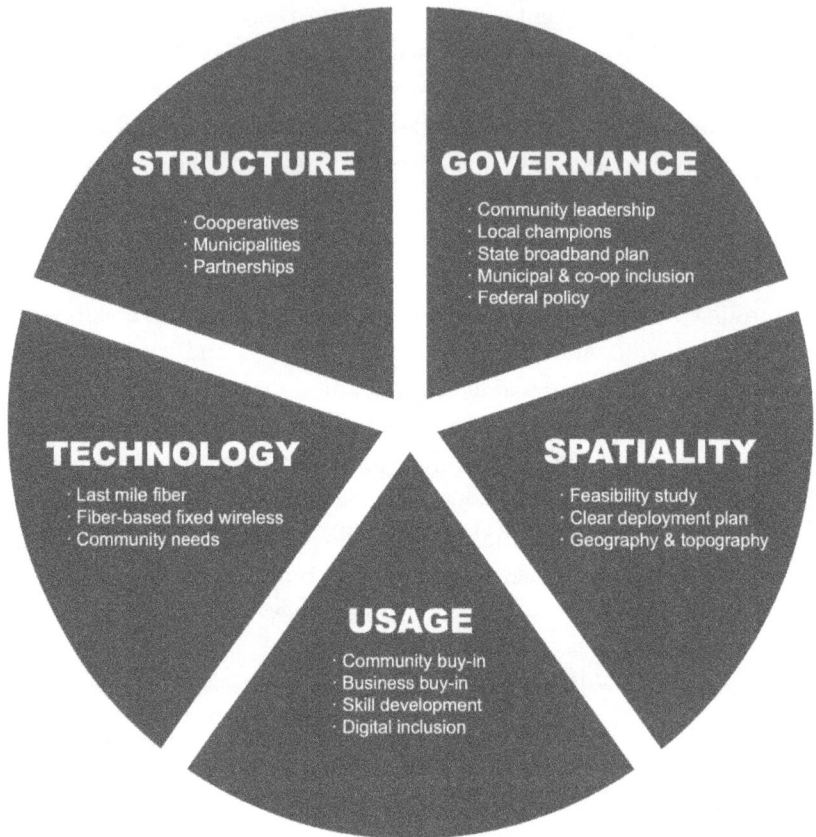

Figure 4.5
The rural broadband model.

As Rock County demonstrates, each element in the model needs to be considered alongside and in tandem with the other elements. Rock County was set on fiber for its technology, partnered with Alliance Communications for a hybrid public-private partnership, had strong local digital champions, tapped into Minnesota's broadband infrastructure, and is now piloting programs for digital inclusion. What the rural broadband model demonstrates, therefore, is that while there are no cookie-cutter solutions to efficient rural broadband deployment, *localism* is the decisive component.

Rock County is not alone in its broadband successes. Across the state and across the country, communities have come together to solve their digital dilemmas and break down their broadband barriers. They have been more successful than the major telecommunications companies at connecting their communities, educating their populations, and keeping prices affordable. None of this is possible without local considerations of spatiality, technology, usage, structure, and governance. States like Minnesota and organizations like the Blandin Foundation are key to the success of rural broadband projects. Fortunately, Minnesota is not unique. Washington State has empowered local port authorities to provide broadband connections, South Dakota and North Dakota both have powerful statewide fiber networks, Missouri has just launched its rural broadband office, and Georgia and Tennessee are breaking down legislative barriers to cooperative broadband. Solving the digital divide requires a multistakeholder "all-hands-on-deck" approach, fostered by robust federal policy, coordinated by state policy, and championed by local communities. As we saw above, Nicholas (2003) thinks of geography as a "geo-policy barrier" for rural telecommunications. He is concerned with the ghettoization of rural communities and the isolation of remote communities. While I share the concern, what Rock County demonstrates is that rurality can be a geo-policy *opportunity*. Its story reminds us the solution to rural broadband deployment is not always to incentivize large telecommunications companies to altruistically enter rural markets (as FCC policies tend to do), but rather to empower communities to create those markets for themselves. For this to happen, however, we need a comprehensive understanding of where broadband deserts (areas without connectivity) and broadband droughts (areas with poor connectivity) exist. For the Midwest, this poses a particular challenge because of the vast amount of land devoted to agriculture. How cropland should be treated in rural broadband policy is the subject of chapter 5, this book's penultimate investigation of the future of rural connectivity.

5   The Jolly Green Giant Goes Digital: Broadband to the Farm and Precision Agriculture

If you haven't been to Blue Earth, Minnesota, I suggest a visit as part of your next road trip. This city of a little over 3,000 is located in the middle of southern Minnesota, some 130 miles south of Minneapolis and 190 miles north of Des Moines, Iowa. As its name promises, Blue Earth is nestled firmly in the lush fields of America's grain belt (notably corn and soy), which makes for a stunning, if not hypnotic, drive.

Blue Earth's most famous resident is a fifty-foot statue of the Jolly Green Giant—a "can't miss" attraction. This friendly colossus who reminds us to eat our vegetables found a home in the town because of a local canning facility and, allegedly, after a local radio station owner gave out Green Giant vegetables to passersby. It was hoped the statue, erected in 1979, would continue to bring visitors to Blue Earth after the construction of Interstate 90 bypassed the town. To a certain extent it worked, and some 10,000 people visit Blue Earth every year.

Blue Earth is also home to Bevcomm, a 124-year-old, family-run telecommunications company serving the town and surrounding area. Bevcomm is what brought my hound dog, Tuna, and me to Blue Earth on our rural broadband road trip across the Midwest. There, we met with the company's owner and general manager Bill Eckles, who took us to some of the more remote places Bevcomm serves to teach us about broadband to the farm. Our first stop was a hog and soybean farm outside of Wells, Minnesota. Here, I met a young farmer who paid for a dedicated connection to his farm and established his own fixed wireless network. Through this network, he monitors all 2,500 of his hogs on his iPad, which includes notifications on how much water they drink (an indicator of illness, I was told), along with food consumption and movement. The network also allows him to implement the latest generation of what is called "precision agriculture" equipment on his adjacent soybean farm. Precision agriculture is the application of ICT to the practice of agriculture (Shannon, Clay, and Newell 2018; Shimmelpfennig 2016; Whitacre, Mark, and Griffin 2014). This has also spawned terms like "smart farming"—the use of big data to inform growing decisions (Wolfert et al. 2017) and "the connected farm" (Hambly and Chowdury 2018). Sitting in the cab of the young farmer's seeder, I learned that the latest generation of agriculture technology and Global Positioning System (GPS) guidance not only allows precision navigation and automated steering but generates a map of his field based on the previous year's harvest. This allows him to plan next season's crop.

Our next stop was Aker Technologies in nearby Winnebago. Here, I got to see the next generation of agriculture equipment: drones, or unmanned aerial vehicles (UAVs), designed to take high-resolution images of cropland to look for growth patterns, drainage, "spray drift" (the drift of herbicide), soil compaction, and off-season field management. Between on-the-ground GPS, mapping and monitoring technologies, and the UAVs' high-resolution imagery, farming and farmers are entering a new era of agriculture. None of these advances in agriculture would be possible without broadband. As the folks at Aker kept telling me, their business in rural Minnesota would have been impossible without the fiber-optic connection provided by Bevcomm.

This chapter is about the intersection of precision agriculture and rural broadband. It offers a deep dive into one of the most celebrated applications of broadband in rural America. Given the exciting developments in agriculture technologies, it is not surprising that so many have jumped to promote rural broadband in the name of precision agriculture. Indeed, some of rural broadband's most ardent champions have been the agriculture associations of America, namely, the NFU, AFBF, the Association of Equipment Manufacturers, and the Grange. As the AFBF argues,

> Farmers and ranchers rely on broadband access to manage and operate successful businesses, the same as small businesses do in urban and suburban America. Access to broadband is essential for farmers and ranchers to utilize the latest precision agricultural equipment, follow commodity markets, communicate with their customers, gain access to new markets around the world and, increasingly, for regulatory compliance. (Moore 2018)

This industry rhetoric has proven persuasive, with multiple congressional hearings on precision agriculture,[1] and the inclusion of the Precision Agriculture Connectivity Act of 2018 in the omnibus 2018 Farm Bill, which ordered the FCC to establish the Task Force for Reviewing the Connectivity and Technology Needs of Precision Agriculture in the United States.

The future of agriculture depends on broadband, but it is a future in which the political economy of rural broadband is tied to the political economy of agriculture in complex and contradictory ways. The relationship between broadband and agriculture illustrates what Mary Hendrickson, Phillip Howard, and Douglas Constance (2019) call a "mode of power"—the reproduction of power within capitalism—and expands it by looking at how interindustry modes of power serve to reproduce themselves. Based on Howard's (2016) use of the term, Hendrickson, Howard, and Constance (2019) explain that "mode of power . . . does not assume that capitalists are driven to increase production (nor consumption) but only their own power relative to everyone else, even if it reduces well-being" (20). Drawing on this notion of mode of power, I argue that the exuberant discourse of policymakers and industry that link rural broadband and precision agriculture masks deeply entrenched and unresolved political economic issues that span both industries. The broadband-agriculture mode of power manifests in three cases: *representation* on the broadband map, *ownership* of the data generated by precision agriculture practices, and *control* of the technology. Each of these dynamics replicates and reinforces the power of incumbent actors, most notably Deere & Company (more commonly known as "John Deere," and hereafter "Deere"), while disempowering users and farmers. Ultimately, I argue that both broadband enthusiasts and policymakers need to be more critical of these practices, just as we need to be more critical of the promises of the major telecommunications companies to connect rural America. It is easy to get wrapped up in the new and the next, and just as easy to forget the ongoing message of this book—that broadband is about people, not about technologies—and certainly not about companies. Forgetting this fact risks blindly entering telecommunications and industrial agriculture's walled gardens, black boxes, and what Marc Andrejevic (2007) calls "digital enclosures." These refer to instances when our entire digital presence exists within a single company's digital ecosystem. Until these issues are resolved, a more temperate discourse is required so as not to cede rural Americans' digital sovereignty to the major agribusiness and telecommunications companies and further exclude and disenfranchise the farmer and rural broadband consumer.

This chapter embodies a component of critical political economy known as "social totality"—placing the object of study within a larger political economic framework (Mosco 2009). I seek to understand the tensions between the modalities of rural broadband and the realities of contemporary agriculture, which very much requires broadband. Connecting the dots between chapters, we will see that how agricultural companies and organizations discuss and use broadband further exemplifies the politics discussed in chapter 2. As a caveat, I am sensitive to the work coming out of critical rural studies that reminds us not to reduce the rural to the agrarian (Thomas et al. 2013). There is much more to rural America than farms. That said, the fact that agriculture is so often invoked in policy conversations over rural broadband requires us to fully address this connection and contemplate the implications of this oft-invoked relationship. I begin, therefore, with a thorough explanation of precision agriculture and broadband to the farm. I then turn to the three cases of the broadband-agriculture mode of power: representation, control, and ownership. I conclude with thoughts on 5G and the future of connected agriculture.

Precision Agriculture

As noted above, precision agriculture is the application of ICTs to the practice of agriculture (Whitacre, Mark, and Griffin 2014). These technologies range from GPS-embedded tractors, combines, and seeders to barns and pens equipped with sensors for monitoring livestock, from sensors in fields monitoring moisture and pH to drones taking high-resolution images, and from the mapping and monitoring of yields to the ability to prescribe seed plans, fertilizer input, and water. The connected farm is the agrarian manifestation of what Vincent Mosco (2017) has called "the next internet," which is made up of cloud computing, big data, and the IoT.[2] The IoT has become shorthand for Internet Protocol-connected or "smart" devices, ranging from watches to cars and from refrigerators to tractors (Cirani et al. 2018). Richard Adler (2015) calls the intersection of agriculture and broadband "the internet of *growing* things." He writes, "Together IoT sensors and big data are beginning to supplant the Farmers' Almanac as the indispensable information resource for modern agriculture."

Precision agriculture allows the farmer to approach the field, or livestock, segmentally, rather than as a whole. Segmentation, in turn, allows the farmer to get the most out of their land, thus improving yield, soil ecology, and efficiency. The USDA estimates that adoption of precision agriculture "can produce a 3–18 percent boost in crop yields" (Perdue 2017, 31). Precision agriculture has been around since the 1990s, when large farm equipment started to be equipped with GPS devices that informed a base station of their location (Gibson 2019). Since then, GPS has remained central to precision agriculture, but the ecosystem has expanded dramatically to include a suite of three interconnected practices that I place under the headings of mapping, monitoring, and management.

Mapping

Mapping is the first and foremost element of precision agriculture and includes GPS, geographic information system (GIS) mapping, and automated steering. GPS allows farmers and technicians to locate a tractor on a piece of land anywhere in the world. It also requires relatively little bandwidth for transmission (around a 2G signal). GPS begat the field of telematics, which is defined as the "transmitting of data through wireless communication links between the home base and in field units" (Whitacre, Mark, and Griffin 2014, 2). Today, not only is the location of the vehicle transmitted, but so is its condition (Whitacre, Mark, and Griffin 2014). Deere's JDLink system, to use but one example, connects equipment to Deere's network and the grower (and back), transmitting information on location, idle time, fuel usage, and maintenance. Remote display access—which is basically a PC in the cab— allows the driver to access this information. Starting in 2006, all of Deere's heavy equipment was connected via telematics, a practice that extended to tractors, sprayers, harvesters, and combines in 2011 (Deere 2014b).

GPS also permits the automation of certain aspects of the agriculture process, most notably precision guidance and automated steering. Humans make mistakes, with one early study finding that conventional row crop operations lose upward of 10 percent of their field to human error (McKinion 2010). Conversely, "signals from the auto-steer system allow tractors, sprayers, combines and harvest equipment to navigate a predetermined path with centimeter accuracy. Auto-steer systems allow extended hours of operation by operators without the fatigue associated with non-auto-steer vehicles" (McKinion 2010, 837). This allows farmers to maximize their acreage by reducing human error and to save on seed usage (with seeds you pay for what you use) and ecology ("where the plants would compete against each other" because of double planting) (Greene 2016). Guidance systems are also useful for the application of fertilizer, which can reduce the amount of chemicals being introduced into the ground. When coupled with M2M technology, auto steering and telematics can also wirelessly tether one vehicle to another, allowing a harvester to be linked with a tractor pulling a grain cart. This near-perfect synchronization allows the combine to deposit its yield in the cart without waste.

As fascinating as telematics are, the intersection of GPS and *mapping* is where maximal use value is located. Because we know where the tractor is, its movements can be mapped via GPS. Combining GPS location data with monitoring and sensing data (discussed below), GIS is used to create digital maps that can depict anything from yields to soil moisture to pH balance. These maps and plans are loaded on to the tractor's onboard computer before the necessary agricultural task (tilling, seeding, fertilizing, harvesting) begins. More advanced technologies coupled with advanced broadband networks even allow certain tasks to occur in real time, therein eliminating the temporal gap between data gathering and decision implementation (Tzach 2018).

This next generation of real-time precision agriculture is only possible with high-speed broadband networks, which are uncommon on North American farms (Hambly and Chowdury 2018; Whitacre, Mark, and Griffin 2014). Helen Hambly and Mamun Chowdury (2018) note that broadband availability is a major predictor of precision agriculture adoption, and their study of farmers in Ontario, Canada, found that only 3 percent of farmers may have the broadband connection necessary to take advantage of real-time big data transfer. According to the USDA's 2019 census, only 75 percent of American farms have a computer with an internet connection. Of these, 3 percent use dial-up, 22 percent use DSL, and 26 percent use satellite (see table 5.1). In sum, over 50 percent of American farms with an internet connection have an inadequate connection and are unable to access many of the mapping technologies discussed in this section.

Monitoring

Yield monitoring is one of the most widely adopted precision agriculture technologies for row crops. A 2016 USDA study found that 40 percent of large corn farms, representing 70 percent of all corn cropland, used yield monitors, while a quarter of farms translated that data into yield maps (Schimmelpfennig 2016). Similar numbers are present for soybean farmers. Most combines come with yield monitors, a series of sensors located throughout the machine.[3] As Schimmelpfennig (2016) explains,

> For farm yield data, harvester-mounted yield monitors gather data in the grain elevator of the combine. As paddles in the elevator rotate and eject the harvested corn to waiting truck, the train bounces off a load cell that measure the corn mass-flow that is converted to an electrical signal captured on a flash drive in the chute. (3)

The data is then transmitted—either digitally or manually (depending on broadband connection)—back to the home computer. When combined with GIS mapping technology, a color-coded map of the area is created. For instance, a soybean farmer in Minnesota uses color-coded maps to plant her crop based on last year's yields. Green indicates a good haul, yellow a weaker haul, and red the worst. This data helps determine how much seed to plant where and how much fertilizer to apply.

Soil monitoring is another way to mobilize precision agriculture and can be accomplished either by manually extracting a portion of the soil and testing it, by directly embedding sensors in the soil, or by UAVs. Sensors test for soil type, moisture levels, nitrate levels, and pH level. Data for soil maps can also be found from public sources, most notably the USDA, which maintains a large database of aerial imagery. Like yield monitoring, soil is analyzed, the results are GIS mapped, and then a plan is developed for seeding, watering, and fertilizing. The entire process may take days or weeks. That said, the technology exists for this information to be transmitted wirelessly, with the newest generation of soil monitors equipped with modems (Tzach 2018). As we saw above, Aker Technologies in Winnebago, Minnesota, has developed drones for soil mapping and sensors to measure and monitor pest infiltration and blight. Aker uploads gigabytes of data daily from their drone flights. One 430-acre farm, for instance, generated 12 GB worth of data. As was said to me again and again during my visit to Aker, nothing short of a fiber-optic connection can handle these giant data streams. High-performance broadband is as necessary for modern agriculture as it is for any tech startup or Silicon Valley incumbent.

Management

Combining mapping and monitoring allows farmers, or, more often than not, third-party data processors, to develop prescriptive action plans for farms (the "smart" in smart farming). Indeed, the practice of data management and precision agriculture has spawned an entire cottage industry of third-party data processors eager to relieve farmers of their raw data and translate them into actionable plans. Key here is a practice known as variable rate application (VRA), which is the "adjustment of the amount of cropping inputs such as seed, fertilizer and pesticides to match conditions in a field" (Shannon, Claw, and Kitchen 2019, 255). Recall the yield maps I mentioned in the previous section: in the green section the farmer may want to plant more, confident the soil can handle more seeds; the yellow and red areas might require less seed, more deeply planted seeds, or more attention throughout the growing season (Schimmelpfennig 2016). Thanks to GPS and telematics, much of this variable rate application is done automatically, without much input from the driver. As Jane Gibson (2019) explains, "Precision technologies can generate data on topography, soil nutrients, moisture, pH, tilth, root-zone capacity, soil compaction, yield variability . . . that can be mapped, measured, and analyzed for optimal input prescription to improve yields for particular areas of a farm and reduce input costs by minimizing waste" (144). Technology even exists to plant different types of seed in different locations "with a single pass of the tractor" through what is known as "precision planting" (Schimmelpfennig 2016, 4). Variable rate technologies (VRT) are the slowest to be adopted among large corn and soybean farms, with 19 percent and 26 percent respective adoption (Schimmelpfennig 2016). Still, the trend is for greater adoption over time.

The practice of management also encompasses home computer use. Digital farm home management is necessary for market assessment and coordination, along with the buying and selling of equipment, crops, and livestock. As a recent USDA (2019a) study attests, "Web platforms connect farmers directly to buyers, allowing them to earn premiums for meeting specific quality standards and bringing between $0.35 to $0.51 more per bushel for corn, soy, wheat and rice" (26). Even for those who have yet to adopt precision agriculture technologies, farm computer usage is essential. Over 50 percent of farm family income is derived from nonfarm-related activities, and many of these are home-based businesses requiring a robust broadband connection (CRS 2019b). As a representative from the NFU explained to me,

> There are very, very few local farmers who, in many ways, use a lot of physical labor. Their farming operation isn't necessarily extremely technologically advanced, but they have to have a Facebook page, they have to have a website, they have to have social media to really access markets and promote their products. (personal communication, 2/14/18)

I interviewed a soybean farmer from Kansas who used the example of banking:

> I live about 20 miles from my bank, and it just makes a lot more sense to do banking on the computer rather than having to pick up the phone or drive all the way out there to talk to somebody. . . . And then, obviously, we have an international market for our grain so it helps having access to . . . real-time data about where the markets are going, at our fingertips. (personal communication, 8/13/19)

Management, therefore, applies not only to the land but to the entire operation of contemporary agriculture. As the next section describes in detail, despite the importance of both home computers and high-performance broadband to modern agriculture, American farms remain woefully underconnected.

Broadband to the Farm

The USDA conducts a census of American farmers' home computer use every two years. Its latest report, released in August 2019, found that only 73 percent of farmers have access to a desktop or laptop (69 percent own a desktop or laptop), while tablet and smartphone usage stands at 52 percent. As noted previously, only 75 percent of farmers have access to the internet, with satellite and DSL connections most common and dial-up a continued presence (see table 5.1).

The majority of these connections (specifically, dial-up, DSL, and satellite) are subpar for everyday life, let alone the requirements of next-generation agriculture such as those produced by Aker. The figures are also incomplete. A recent study by the United Soybean Board (2019) found that upward of 60 percent of farmers say their internet connections are inadequate to run

Table 5.1
Primary method of internet access on American farms (2017 vs. 2019)

	2017	2019
Dial-up	3%	3%
DSL	28%	22%
Cable	16%	16%
Fiber	9%	12%
Mobile	19%	18%
Satellite	23%	26%
Other/unknown	2%	3%

Source: USDA (2019b).

their business, 78 percent do not have a choice in service provider, only 40 percent of respondents had a fixed connection, and 37 percent want to increase their use of data in their operations. After noting that "$13 billion in annual farm equipment purchases are impacted by lack of rural internet," the report concludes with a familiar refrain: "Fast, reliable connectivity will be the linchpin in the American farmers' future success" (United Soybean Board 2019, 11).

While basic precision agriculture, specifically GPS location, requires little bandwidth, more information-heavy activities like yield monitoring, soil monitoring, automation, and aerial drone imagery require massive levels of bandwidth. Take a 2014 study by Shearer as reported by Tyler Mark, Terry Griffin, and Brian Whitacre (2016):

> Shearer (2014) estimated that row crop producers potentially generate 0.5 kilobytes of data per plant. In other words, a corn producer with a plant population of 30,000/ac could produce 15 megabytes of data per acre; and if this were a 1,000 acre corn farm, they could potentially produce 15 gigabytes of data per year that would need to be transferred. Extrapolating this out to the approximately 88.9 million acres of corn planted in 2015 there would be approximately 1,333.5 terabytes of data produced. This estimate does not include the usage of drone or UAV imagery data that is increasing in popularity in agriculture; the reliance on imagery from drones greatly increase these data transfer requirements. The amount of data generated from drones will depend on the type, frequency, and quality of images that are being taken (Buschermohle 2014). (51–52)

The geographically vast nature of agriculture renders wireline connections unfeasible. This means that farms are dependent upon mobile, fixed wireless networks, and/or satellite for their connections. Hambly and Chowdury (2018) report that while the basic precision agriculture applications can be managed with a 2G or 3G mobile signal, more advanced actions like device monitoring and remote control require a 4G signal with between 1.5 Mbps and 5 Mbps available. As Mark Lewellen, manager of spectrum advocacy for Deere, said in an interview with me, "We need basically LTE type speeds . . . but coverage is a problem for us. We don't have cell coverage out in America's cropland" (personal communication, 7/25/17). The FCC's 2019 broadband deployment report found that over 30 percent of rural Americans lack an LTE connection at 10/1 (FCC 2019a). Worse, a separate report by the FCC (2019d) found that mobile providers exaggerated their levels of connectivity by almost 40 percent. This lack of connectivity costs farmers both time and resources. It means the difference between wireless transmission and having to save data on a USB drive and manually transfer it to a home computer, or, worse, having to *drive* that USB drive to a nearby town for upload.

The poor level of connectivity on American farms has united policymakers, companies like Deere, and farming associations like the AFBF, NFU, and the Grange. Each of these actors has championed rural broadband for agriculture. In 2016, for instance, the Senate Subcommittee on Communications, Technology, Innovation and the Internet heard from Darrington Seward, managing partner of Seward & Son Planting Company, who said,

> Our main goal in precision agriculture is to farm as many acres as we can while minimizing inputs and increasing our yields in an environmentally sustainable way. We depend on reliable and speedy broadband connections. Without reliable broadband, our production practices would be completely compromised. We would suffer yield losses, decreased productivity, and reduced profitability in an industry with ever tighter profit margins. (US Congress 2016)

This sentiment was anticipated by Deere in written testimony to the House Committee on Agriculture's hearing on big data in agriculture in 2015:

> It should also be noted that, without essential broadband connectivity to croplands, many of the potential benefits of "big data" in agriculture can never be realized. Real-time ag services using data generated on the farm are dependent on reliable, high-speed wired and wireless connections to the Internet—connections that in turn depend on a robust rural broadband infrastructure that is currently lacking in many parts of the country. (US Congress 2015, 71)

The USDA has also taken up the cause of broadband and precision agriculture:

> E-connectivity is not simply a rural issue; Internet expansion, economic productivity, and food security contribute to each citizen's quality of life, regardless of where they live. The benefits of broadband e-connectivity accrue not only to the producers using Next Generation Precision Agriculture technologies, but also to consumers throughout America and the world who value a safe and efficient food supply. (USDA 2019b, 41)

Perusing these examples, it is easy to see how one can be exuberant about the possibilities of precision agriculture. Enthusiasts champion its ability to reduce environmental damage, increase yields, and raise efficiency. This is an important line of argumentation. But we also need to be mindful of who is making these arguments, how these actors stand to benefit, and what is *not* being debated. To reiterate, I argue that the fight for rural broadband in the name of precision agriculture is a fight over representation on the national broadband map, ownership of farm data, and control of agricultural technology. Said differently, as much as corporations argue for the need to connect American farms, they are also arguing for the power to populate the broadband map, own farm data, and control agricultural technology. The next section evaluates each of these claims, beginning with representation.

Representation (on the Map)

The issue of representation manifests in debates over the national broadband map, which were discussed in detail in chapter 2. Recall that the current iteration of the map as managed by the FCC and the data points that populate it are deeply flawed. More than flawed, the map dramatically privileges the voices of major telecommunications companies who can boast of universal coverage in marketing campaigns because of the reporting criteria of the FCC's Form 477. Erroneous data has negatively impacted rural communities, who often find themselves unconnected in practice but "connected" on the map, leading to exclusion from federal grants and dismissal from the policy conversation. Improving and correcting the map has therefore been a key priority of rural broadband advocates, communities, and small ISPs.

The national broadband map, like all cartographic projects, is a contested attempt at representation. There is considerable power in deciding not only *what* gets counted as a data point but also *who* gets counted (Bargues-Pedreny, Chandler, and Simon 2019; Crampton 2004). As cartographer Robert Williams wrote in a 1966 letter,

> Since maps are perhaps the most persuasive form of communication, those who make them must accept an unusually high degree of responsibility of their truthfulness. This responsibility is increased when the aura of infallibility of the computer is added to the map. (qtd. in Wilson 2017, 61)

Whoever controls the map controls the power of representation and has the *perception* of truth on their side. "As mapping platforms often pre-determine places, and their meanings, they shape users' spatial imaginations and limit what is possible to map" (Specht and Feigenbaum 2019, 47). For broadband this can mean determining what areas get funding or are left out, what technologies count as broadband, and what areas are prioritized.

In conversations about redrawing the national broadband map to better reflect the lived realities of American communities, Deere has emerged as a vocal stakeholder. Deere is the largest producer of agricultural equipment in the world, with a North American market share of 53 percent and annual global revenues of $33.3 billion (Reibel 2018; Tita 2019). Over the last twenty years the company has pivoted toward precision agriculture, beginning in 2002 when it introduced JDLink, the telematics-fueled software to locate equipment through GPS (Hest 2010). Incidentally, 2002 is also when Deere developed its automated steering. Today, through multiple proprietary systems, including JDLink (remote connection to machine status), StarFire (GPS), and Machine Sync (M2M telematics), Deere is thoroughly invested in the "internet of growing things" (Adler 2015). As such, it is also fully integrated into America's rural broadband ecosystem. It has a long-standing relationship with AT&T and uses the telecommunications giant's "LTE network for the majority of its connections" (Marek 2018). As part of this ecosystem, Deere has taken note of the rural-urban digital divide, admitting that only 70 percent of the data transmitted through its technology is transmitted successfully because of poor rural connectivity (Deere 2015b).[4] Examples such as this have formed the basis of a decade's worth of Deere's filings to the FCC on the topic of rural broadband mapping and funding. Deere wants the FCC to consider three recommendations:

1. Recognize croplands as areas that require broadband.
2. Map cropland broadband deployment.
3. Count farms as anchor institutions.

Since 2014 Deere and the trade association it leads—the Agricultural Broadband Coalition—have lobbied the FCC to both discursively and politically recognize the notion of croplands as a viable locale for broadband deployment.

> By reviewing those areas of the economy that lack broadband access, rather than simply focusing on population-based coverage, Deere hopes that the Commission can start closing the broadband gap in rural and other underserved areas. In this regard, Deere urges the FCC to examine "cropland" coverage as a key indicator of where deployment gaps exist. (Deere 2015a, i)

Deere wants the FCC to take into account the number of *modems* present on croplands since almost all new large farm equipment comes with a modem, and modems have been installed on older equipment. In its words, "This step would allow machine-to-machine mobile broadband transmissions by agricultural equipment . . . to be counted in the justification for broadband expansion" (Deere 2014a, 7). It also wants croplands included in the FCC's definition of unserved and underserved areas:

> It is unclear how advantageous Form 477 data will be to determine service availability on rural roads and cropland given that information is collected based on fixed and permanent customer locations. The Commission should count machine-to-machine mobile broadband transmissions by agricultural equipment in the field and associated operators' mobile devices when assessing the status of mobile deployment. By counting the number of machines with modems working the 350 million acres of cropland in the United States, the Commission will have better information to more accurately assess the need for advanced broadband services in rural areas and to consider ways to strengthen funding to those rural areas of the country that need it most. (Deere 2015b, 4)

The cropland recognition argument has proven successful, with twenty-four members of the Senate signing a letter petitioning the FCC to "consider a metric of broadband access in croplands (and farm buildings), or some other geographic measurement, in addition to road miles, to identify these areas of greatest need" (Wicker et al. 2016, 2). This language is almost verbatim that of Deere and the Agricultural Broadband Coalition.[5]

Through its interventions to both Congress and the FCC, Deere is attempting to hijack the policy debate over rural broadband and steer it towards its own interests and against the public interest. These actions reinforce my larger argument that, to its detriment, US broadband policy is not about people; it is about technology, politics, and profits. Connected farms mean more connections to Deere's network of proprietary systems. This is particularly true of Deere's push to have farms counted as anchor institutions:

> But unlike community centers, healthcare centers, schools, libraries, scientific institutions, and other similar enterprises that can (and do) build where broadband services are available, agriculture happens where the farmland is. Farms cannot relocate operations in order to follow the broadband. Thus, unlike many other important types of American institutions, agricultural operations are structurally most at risk of being left behind the broadband economy. (Deere 2015a, 12)

Presumably, Deere made such an argument to have farms gain access to other segments of the USF (see chapter 2), most notably E-Rate. Deere (2015b, 4) suggested that "support should be available for the deployment of middle-mile facilities and fiber backhaul, which are critical to expanding both fixed and mobile services in rural agricultural centers." The company's logic is as such: increased funding for connected farms means more customers for its digital services. The FCC was not amenable to this request and has not included farms as part of the eligible set of institutions for USF funding.

In its lobbying efforts to the FCC in both mapping and funding, Deere is flexing its mode of power, illustrating an attempt to reify its position as America's dominant agricultural company. For Deere, however, representation of the map and *on* the map is only one part of its plan; ownership of agriculture data and control of precision agriculture technology are the other two components.

Ownership (of the Data)

With the rise of smart technologies that produce tremendous amounts of data, ownership over that data has become a major public policy issue (Draper 2018; Kichin 2014; Perzanowski and Schultz 2016). We need to be concerned, critics tell us, when we surrender ownership over the IoT. Aaron Perzanowski and Jason Schultz (2016), for instance, offer a cautionary tale about what happens when the ownership of digital content is ambiguous, citing deleted e-books on Amazon and cantankerous Keurig machines that only use Keurig products. A parallel set of questions exists over the ownership of the data we produce in our digital wanderings. Since the 2018 Cambridge Analytica scandal that saw the profiles and data of millions of Facebook members collected and used for political advertising without their consent (Vaidhyanathan 2019), and even before with concerns over behavioral targeted advertising (Turow 2013), policymakers have given increased attention to who owns the data produced by our digital lives. As Nora Draper (2018) explains, "The twin issues of privacy and data ownership have become contentious in the digital moment in part because the terms governing information collection and use conflict with the relationship we instinctively feel with our personal information" (173).

These concerns are echoed in agriculture, where the practice of precision agriculture, as noted above, generates a considerable amount of data. According to the Association of Equipment Manufacturers (2017), one ear of corn produces approximately 1 KB of data per year. When multiplied by the number of plants in a field and the number of fields across the country, we're talking about a tremendous amount of data:

> When you consider that a typical corn field has about 32,000 plants, that means that every corn field is producing about 32 megabytes of data annually—or about as much bandwidth as it will take you to stream the first four singles off Taylor Swift's new album. But with 450 million acres of corn in the United States, that means that farmers are already producing 14.4 million gigabytes of data related to their corn crops annually—nearly 10 times as much bandwidth as has been used to stream Taylor Swift's latest single 205 million times this year on Spotify. (Association of Equipment Manufacturers 2017)

In most instances, the raw data produced by crops and equipment is transmitted (either wirelessly in real time or with a lag for physical movement) to a data-processing company to clean it up and produce a detailed planting plan. Deere, for instance, offers its own data-processing service, but alongside it has grown a cottage industry of precision agriculture consultants relying on precision agriculture data. Agris Co-operative (2019), from Ontario, boasts "farm data so local, it's personalized."

"Immanent commodification" describes the process of one commodity spontaneously begetting another (Mosco 2009). Agriculture data has fostered agriculture data analytics, joining the likes of audience ratings as an immanent commodity. "Data," as Rob Kitchin (2014) argues, "can thus be understood as an agent of capital interests" (16). Here, the commodification of agricultural data is no different than the commodification of data in any other industry. This includes both the market celebration over the growth of a nascent industry and the growing concerns for privacy and ownership. As Vincent Mosco (2009) explains,

> Internet cookies, digital television recording devices, "smart" cards, etc., produce new products, in the form of reports on viewing and shopping, containing demographic details that are linked to numerous databases. But these new products are more than discrete units. They are part of a commodification process that connects them in a structured hierarchy. The implications for privacy are powerful. (143)

Data collection has become a major focus in all facets of agriculture, from equipment to seeds to markets (Bloch 2018). Deere, Bayer (which bought Monsanto in 2018), and Dow DuPont (now Corteva Agriscience) have all made major investments in data-gathering techniques and companies. In 2014, for instance, DuPont predicted that it would generate $500 million annually from its data services division (Bunge 2014). A Monsanto-Bayer subsidiary, the Climate Corporation, offers an app called Climate FieldView to

> help farmers get their data in one place, uncover valuable field insights, and optimize their inputs. . . . FieldView uses field, planting and yield data to help farmers gain a better understanding of their fields, and seed and input performance to help them make data-driven decisions to maximize their return on every acre. (Monsanto 2018)

As with the data generated by our Instagram posts (see Vaidhyanathan 2019), however, growers are concerned about who owns their data, the interoperability of data from different machines, and the larger implications of agriculture data ownership. "The trouble is," writes Sam Bloch (2014) for *New Food Economy*, "we don't really know what these companies are doing with all the data the tools are hoovering up." Part of the issue is about trust and another part about ambiguous contracts. Farm contracts around data mimic the terms of service one agrees to on social media—they are pages of legal details that many farmers do not read or understand. A 2019 study of Australian farmers, for instance, found that "74 percent of respondents did not know much about the terms and conditions relating to data collection in their agreement with service providers" (Wiseman et al. 2019, 3). As Todd Janzen, cofounder of Janzen Agricultural Law LLC, explained in an interview with *Farmtario*:

> There are companies out there that say "yes, you own the data," but when you read the agreements you find out that they have an unlimited licence to do whatever they want with the data. They own it from the standpoint that they can do whatever they want with it. (McIntosh 2018)

Not knowing breeds mistrust, a finding from Wiseman et al. (2019), who argue that "farmers currently lack trust in the way in which their farm data is being collected and managed" (6).

There are two concerns: what happens if the data is shared with competitors and what happens if the data is used to shape the market. To begin this discussion, here is an excerpt from my conversation with two members of the NFU, its president Roger Johnson and government relations representative Matt Perdue, in February 2018:

Christopher Ali: Is there a concern amongst your members over data privacy and ownership?

Roger Johnson: There's a lot of concern over that stuff, yeah.

Christopher Ali: There is, eh?

Roger Johnson: Yeah. There is. Of course. I mean, there's sort of an innate suspicion in all of us that we don't want folks to know more than what they need to know about what our business is. So, of course, the more this big data becomes a fact of life, the more serious the question about to whose benefit will it be. So yeah, there's a huge concern about that.

Matt Perdue: And I think one of the biggest questions that we have conversations around now is, who owns the data?

Christopher Ali: Yeah.

Matt Perdue: You know, if it's a John Deere system, does John Deere own that data? And that's really what scares farmers, from a variety of different ways. So I do think that's going to be one of the regulatory challenges that we see faced down the line.

Johnson and Perdue bring up the threat of undetermined data ownership along with the difference between ownership and control. Farmers may legally own their data, but the data may be controlled by another company. For a more specific example, we turn to the AFBF, which has been particularly vocal on the issue of farm data ownership. In a 2014 interview with NPR, Mary Kay Thatcher, then senior director for congressional relations at the AFBF, discussed the concerns of farm data falling into the hands of a competitor seeking to rent the same land, or to major companies who can potentially manipulate the market.

> For instance: Your local seed salesman might get the data, and he may also be a farmer—and thus your competitor, bidding against you for land that you both want to rent. "All of a sudden he's got a whole lot of information about your capabilities," Thatcher says.
>
> Or consider this: Companies that are collecting these data may be able to see how much grain is being harvested, minute by minute, from tens of thousands of fields. That's valuable information, Thatcher says. "They could actually manipulate the market with it. They only have to know the information about what is actually happening with harvest minutes before somebody else knows it. I'm not saying they will," says Thatcher. "Just a concern." (Charles 2014)

Deere and the rest of the major precision agriculture companies promise not to abuse the data that is collected (Charles 2014; Rotz et al. 2019). Indeed, all of the major equipment, seed, and agrochemical companies, including AGCO, CNH, Dow, DuPont, Deere, and Monsanto, signed on to the AFBF's 2016 Privacy and Security Principles of Farm Data, which states a belief in data sovereignty:

> *Ownership*: We believe farmers own information generated on their farming operations. However, it is the responsibility of the farmer to agree upon data use and sharing with the other stakeholders with an economic interest, such as the tenant, landowner, cooperative, owner of the precision agriculture system hardware, and/or ATP [Agriculture Technology Provider] etc. The farmer contracting with the ATP is responsible for ensuring that only the data they own or have permission to use is included in the account with the ATP. (AFBF 2016a)

On the one hand, the AFBF wants farmers to have greater control over farm data; on the other hand, it believes responsibility falls squarely on farmers. This goes against their own polling data, which found that 77 percent of respondents were "concerned about which entities can access their farm data and whether it could be used for regulatory purposes" (AFBF 2016b). This position, however, aligns with industrial agriculture's mode of power that reifies its position. Here it is the farmer and not the company that bears the brunt of data responsibility.

Concerns over trust and skepticism over data ownership are well placed. In 2016 Deere tried to acquire precision agriculture company Precision Planting from Monsanto for $190 million. The merger was ultimately blocked by the Department of Justice because it would give Deere an 86 percent share of the high-speed precision planting market.[6] What was not noticed was that Deere and Precision Planting had entered into a data-sharing arrangement in 2015 "that provides The Climate Corporation [also owned by Monsanto] near real-time access to data collected from Deere equipment. The data-sharing agreement offers Monsanto considerable value in addition to the $190 million that it will receive from Deere" (*United States of America* v. Deere & Company 2016, 4; see also Janzen 2017). The lesson here is that, just as with social media terms of service, farmers must be cautious of those in precision agriculture (Janzen 2017).

Despite the promises made by Deere and others, or perhaps because of them, the Pennsylvania Farm Bureau doubled down on its concerns and released a brochure in 2014, which lists the questions farmers should ask before they sign a contract involving data (see figure 5.1). The nine issues to ponder include:

1) Do I own the data?
2) How will my data be used and what benefits can I expect from allowing a provider to include my data in a "big data" database?
3) Will I control the management of that data?
4) What is aggregation of data? How does aggregation protect me?
5) Is my personal farm data anonymized? How is anonymize defined?
6) Can I stop sharing my data once I have agreed to share?
7) Who else might have access to my data? Can it be released to the public or to a third party?
8) What is the value of my data to *me*?
9) What is the value of my data to *the company*?

As the above examples attest, the AFBF clearly sees data ownership as a concern. But frustratingly, they caveat this concern by putting the onus

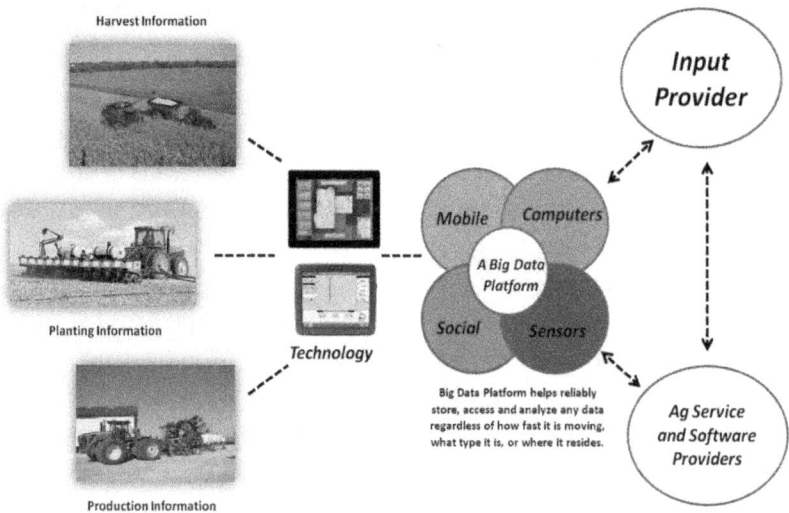

Ponder These Nine ...

Before You Sign

DATA PRIVACY EXPECTATION GUIDE

Harvest Information

Planting Information

Production Information

Technology

Mobile · Computers

A Big Data Platform

Social · Sensors

Big Data Platform helps reliably store, access and analyze any data regardless of how fast it is moving, what type it is, or where it resides.

Input Provider

Ag Service and Software Providers

Pennsylvania Farm Bureau
www.pfb.com • 717.761.2740

Figure 5.1
"Ponder These Nine before You Sign" brochure.
Source: Pennsylvania Farm Bureau (2014).

on the consumer, doubling down on personal responsibility over industry-wide regulations and practices. This neoliberal or "corporate libertarian" approach (Pickard 2013) clouds the gravity of the situation, for it only educates the farmer about predatory corporate behavior instead of calling it out and advocating for change.

However, there is a growing movement among an increasingly data-sophisticated agriculture community away from these industrial and commercial third-party data processors to the open-source community through organizations like Open AG Data Alliance (OADA, n.d.). The OADA asserts that farmers should own the "data generated or entered by the farmer, their employees, or by machines performing activities on their farm" (OADA 2016). Interoperability and data portability are also chief concerns, as the OADA believes farmers should be allowed to bring data from different equipment and precision agriculture software together. Joining organizations like OADA is a host of critically minded scholars and journalists who have taken heed of these issues and been writing vociferously on the topic (e.g., Koebler 2017, 2018a, 2018b; Rotz et al. 2019; Wiens and Chamberlain 2018).

The commodification of agriculture data is another example of the agriculture-broadband mode of power. Ambiguous data contracts cede power to agriculture data-processing companies, including the largest manufacturers like Deere and the largest seed and fertilizer companies like Bayer and Corteva.[7] While the AFBF may equivocate in admonishing industry for abusing its position, they are nevertheless correct that farmers need to seriously consider the data implications of precision agriculture.

Control (of the Technology)

The third and final way the agricultural-broadband mode of power operates is through attempts to control the technology, powered by broadband, that is designed to bring agriculture into the digital age. Today, almost all agriculture equipment and machinery comes embedded with a modem and onboard software. This includes everything from combines, seeders, and fertilizers to ground sensors, livestock pens, and milkers. While the previous section raised concerns over the ownership of the massive amounts of data produced by smart farming, this section steps back and asks: Who controls the technology itself? The simple answer is that, more often than not, control resides with the manufacturers of the equipment, not the farmers who purchase it. Because of the amount of software that permeates farm equipment, manufacturers like Deere have cracked down on farmers' ability to repair their own equipment. These manufacturers have suggested that farmers do not own their equipment but are merely renting it from them (Koebler 2017, 2018a, 2018b; Wiens 2015a). They have thus joined the ranks of companies like Apple, Audi, and AT&T who want to limit repairs of their products to authorized dealers only. But, as in the days of old when farmers mobilized for collective action, farmers are fighting back.

The issue comes down to ownership and control. When a farmer buys a planter from Deere, such as a new 1775NT 16Row30 Drawn Planter, it will set them back roughly $230,000. Despite this incredible sum, farmers are finding it difficult to legally repair their equipment, because Deere is forcing them to only use authorized Deere dealers. Failing to do so risks voiding the warranty on a piece of machinery that costs more than some houses.

In the world of farming, time is money, and money is money. Farmers cannot afford to wait on a Deere repair person and cannot afford to pay repair costs when, in previous generations, they could do it themselves or take it to a trusted local independent repair shop. As one farmer told Jason Koebler (2017),

> You want to replace a transmission and you take it to an independent mechanic—he can put in the new transmission but the tractor can't drive out of the shop. Deere charges $230, plus $130 an hour for a technician to drive out and plug a connector into their USB port to authorize the part.

These actions deny farmers their "right to repair" and the right to ownership. As a farmer in Missouri explained to me,

> I think the biggest anger among farmers is not so much that they can't tinker with their software, because like I said, that's a pretty small percentage of owners who would do that in the first place; it's the fact that they pay hundreds of thousands of dollars for this equipment and here's the manufacturer who sold it to them telling them, "Well, you don't really own all that." (personal communication, 1/19/28)

Deere claims these conditions are in place because they do not want farmers accessing the onboard software, which they argue is protected under copyright law. Deere also claims that "it owns the copies of its code embedded in the tractors it sells to farmers, code that is essential to the functioning of the equipment" (Perzanowski and Schultz 2016, 145). Ownership of code is part of a suite of laws forming digital rights management (DRM), which were enacted as part of the 1998 Digital Millennium Copyright Act (DMCA) and strictly forbid tampering or altering digital software (with certain exceptions to be discussed below). Section 1201 of the DMCA makes it a crime to circumvent digital locks protecting the software without permission. Once a farmer turns on a piece of equipment—say, a combine—that farmer is said to have agreed to its End-user license agreement (EULA), therein agreeing to these draconian DRM terms. As Perzanowski and Schultz (2016) explain,

> By interposing a software layer between farmers and their tractors, John Deere created a practical hurdle. And by wrapping its software controls in DRM, it created a legal one. A quick glance at the John Deere owner's manual gives you a good indication of the result. Almost any problem—from high coolant temperature to a parking break that's not working or a seat that's too firm—ends the same way, with a trip to the John Deere Dealer. (144)

This means more money and more control for Deere.

The utmost fear among farmers is that Deere could remotely shut down their tractors if they are found to have violated these conditions (Koebler 2017; Wanstreet 2018). While Deere's rationale for such measures revolves around curtailing potential illegal actions, like altering carbon emissions, it also spans to the absurd, like listening to pirated music in the cab (Bartholomew 2015; Estes 2015). As Kyle Wiens (2015b), founder of iFixit and a prominent voice in the R2R movement, explained in a *Wired* editorial, "The problem is that farmers are essentially driving around a giant black box outfitted with harvesting blades. Only manufacturers have the keys to those boxes." This is the newest form of digital rent seeking, or what Jathan Sadowski (2020) calls the "Internet of Landlords," where farmers are being denied ownership and instead are merely renting the shell of the tractor.

Farmers have begun fighting back, with tactics like only purchasing equipment made in predigital times (Wiens 2015b) and seeking out bootlegged software from the Ukraine to practice "tractor hacking" (Koebler 2017). Doing so allows them to run the necessary diagnostics and connect a tractor to a computer. A less dramatic tactic has seen farmers join, and now lead, what has become known as the right-to-repair (R2R) movement, which has sought legal remedies to this digital problem.

In 2015, the R2R movement petitioned the US Copyright Office for an exemption in the DMCA that would allow farmers to repair their own tractors by giving them permission to access the onboard software. The office granted their requested exception (Copyright Office 2015). This allowed farmers, along with iPhone owners and car owners, to access the digital software that for years had been denied to them. This was also the ruling that allows cell phone owners to "jailbreak" their phones. Users could not make software modifications that would violate laws (such as emissions standards), but owners could at least repair the equipment. That said, the 2015 provisions only reserved this exemption for the *owners* of the cars and tractors, thereby excluding third-party independent repair shops. This caveat was removed when the exemption was renewed in 2018:

> The expanded exemption enables independent repair shops to circumvent technological protections on the software in the course of diagnosis, repair, or modification of a vehicle function. The expanded exemption thus prevents automobile manufacturers from monopolizing the market for repairing their vehicles. (Band 2018)

The 2018 exemption order also gives owners access to telematics software (Copyright Office 2018). Deere and other car manufacturers were, unsurprisingly, vehemently opposed to this exemption. In response, Deere began instituting repair (or anti-repair) clauses as part of its EULAs. As Koebler (2017), who is one of the foremost journalistic voices in the R2R movement, explains, this new decision shifts the onus from copyright law to contract law: "Violation of the agreement would be considered a breach of contract rather than a federal copyright violation, meaning John Deere would have to sue its own customers if it wants the contract to be enforced." Gay Gordon-Byrne, the president of Repair (the formal organization associated with R2R), told me that the problem is not actually with copyright or patent law but with the EULAs. According to her,

> This copyright thing is a MacGuffin. . . . There is no intersection with patent law. . . . You pay $500,000 for a tractor and you turn it on and you are . . . deemed to have had accepted the end-user's license agreement. Which, by the way, if you try to find it, the only one I've been able to find in that category is John Deere's and if you read the thing, your hair should stand on end because you say, "Wait a minute. I just paid $500,000 for this tractor and as soon as I turn it on, I've agreed that everything electronic in it is not mine." (personal communication, 8/22/19)

Deere has thus implemented technological, regulatory, and now contractual barriers to the ownership of its equipment.

In response, farmers and the R2R movement mobilized again and have taken their actions to individual states in hopes of passing statewide R2R laws that would void prohibitive EULA conditions. While they have not been successful (yet) at passing a law, in 2018 eighteen states considered "fair repair" bills, "which would require manufacturers to sell repair parts and tools to the masses, would require them to make repair manuals available to the public, and would require them to provide circumvention tools for software locks that are specifically designed to prevent third party repair" (Koebler 2018a; see also Gault 2019). Gordon-Byrne told me that up to twenty states are considering R2R legislation. At least two 2020 Democratic presidential candidates - Elizabeth Warren and Barnie Sanders – have weighed in by floating a national R2R law (Gault 2019).

A recent scandal over California's R2R law illustrates, however, that industry is not going down without a fight. In 2018, the Far West Equipment Dealers Association negotiated with the California Farm Bureau Federation (CFBF) to strike a deal whereby equipment manufacturers would provide "access to service manuals, product guides, onboard diagnostics and other information that would help a farmer or rancher to identify or repair problems with the machinery" (Wiens and Chamberlain 2018). But the CFBF "gave up the right to purchase repair parts without going through dealers. [And] farmers can't change engine settings, can't retrofit old equipment with new features, and can't modify their tractors to meet new environmental standards on their own" (Wiens and Chamberlain 2018). In essence, farmers can repair their own equipment but do not have access to parts and diagnostic software to do so. This compromise was in reaction to California's proposed Electronics Right to Repair Act, which was introduced in March 2018 but never made it out of committee. Critics accused the CFBF of bowing to industry pressure, while industry claims the comprise demonstrates their benevolence.

The battle over control of the technologies that fuel our digital lives brings to the fore the mode of power of industrial agriculture—alongside big tech (e.g., Apple), car manufacturers (e.g., GM), and broadband providers (e.g., AT&T)—and highlights the tactics of resistance by American farmers and the R2R movement. That the Copyright Office received 40,000 public comments about the R2R in 2015 speaks to the universality of this issue: Americans want to be able to tinker. That major corporations deny them this ability speaks to the material conditions of technological control.

Representation, ownership, and control represent three dimensions of the agriculture-broadband mode of power. That they remain invisible in the policy conversations over precision agriculture and broadband speaks to the power of industry to shape the debate. This is what critical media scholar Des Freedman (2010) has called "media policy silences." If left unchecked, they will continue to favor the dominant positions of major corporations and render users, farmers, and owners increasingly powerless. As the next section discusses, these issues need to be resolved immediately, before the broadband industry shifts gears entirely to the prospects of 5G deployment.

Conclusion: 5G and the Future of Connected Farms

As discussed in the introduction to this book, 5G is still years away for most urban Americans and remains only a mere possibility for rural and agrarian America. That said, both the agriculture and broadband industries are already hard at work creating 5G hype: "5G . . . may also revolutionize the farming industry, which has long been slow to adopt new innovations," reported *Fortune* magazine in 2020 (Reisinger 2020). "With 5G speeds and bandwidth, more applications are set to emerge taking the IoT in agriculture to new heights," Taiwan-based Lanner Electronics tells readers on its website: "From helping with the irrigation of blueberries in Chile, to helping to fight crop diseases in India, the sheer number of potential applications of IoT technologies in agriculture almost appears limitless" (Lanner 2018). When approaching such rhetorical exaltation, we need to ask: Who benefits from 5G deployment in agrarian America? Again, we look to the same recipients as with previous technologies: major telecommunications companies—in this case, AT&T, Verizon, and T-Mobile—and major equipment manufacturers like Deere. They are the ones who own the spectrum, who can afford to roll out 5G, and who can develop the technologies fast enough to gain first mover advantage (Gagliordi 2018). Deere, for instance, bought Blue River Technology in 2016 to help it harness 5G and the IoT (Ferguson 2018). T-Mobile has made rural 5G deployment the linchpin of its merger with Sprint (Engebretson 2019c).

The stakes are astronomical. It could cost upward of $300 billion to connect the country with 5G (Goldman 2018). The benefits will be reaped by 5G providers, which in 2020 are AT&T, Verizon, and T-Mobile/Sprint. Adding to this, the FCC has pledged to develop a $9 billion fund for rural 5G deployment (FCC 2020e). It is also expected that the IoT will bring worldwide technology spending to an astounding $1.2 trillion by 2022, a boon to technology companies, including industrial agriculture (Columbus 2018).

With these stakes firmly in mind, this chapter introduced us to the concept of the agriculture-broadband mode of power, which describes the ability of these intersecting industries to reify their dominant positions in policy and practice. This chapter has also complicated the rather uncritical way the popular press, industry, and politicians have linked rural broad-band and precision agriculture. To be sure, the potential of these technolo-gies and practices is immense and vital, but we need to ensure they are delivered democratically and in the service of farmers, growers, and con-sumers, not just providers and manufacturers. In rural broadband and pre-cision agriculture there is potential to feed a hungry world—but first we need to ensure that we are not simply feeding a hungry pair of industries. The conclusion to this book offers some thoughts on the democratization of rural broadband deployment, subsidy, and policy, and looks to the future of 5G deployment across rural America.

Conclusion "Everything Is Better with Better Broadband": Toward a National Rural Broadband Plan

In February 2019, I published an op-ed in the *New York Times* that explained some of the conclusions from my research on rural broadband (Ali 2019c). In the weeks following its publication, I was inundated with emails and calls from people across the country interested in sharing their stories about rural broadband. I read and listened to them all and followed up with a few for formal interviews. What I heard reinforced the conclusions I make in this book: rural Americans are frustrated with their broadband service (or lack thereof); incumbent providers are failing rural Americans; DSL and satellite connections are obsolete; current policies privilege the largest companies; grants and loans are too difficult to get; and more needs to be done by our elected officials to connect this country.

I also learned more about the incredible resilience of rural communities in the face of broadband's market and policy failures. I learned about the digital champions in Pomeroy, Washington, who have rallied around a state law that empowers port authorities to fund and operate broadband networks.[1] I learned about how the Missouri Farm Bureau and Janie Dunning brought together stakeholders to rekindle the state's broadband efforts. I learned about Axiom Technology's efforts to bring broadband to the remote islands of Maine. And I learned about a number of counties in my central Virginia backyard looking to better understand the possibilities and opportunities of broadband. These stories reinforced the other set of arguments I make in this book: local broadband is the best broadband; digital champions are crucial to a community's connected success; cooperatives are the unsung heroes of rural broadband; and farms are increasingly relying on high-performance connections. Michael Bell, Sarah Lloyd, and Christine Vatovec (2010) would call these stories "the power of the rural," and I would agree. As they write, "We can and do hear the active rural voice because it is a voice of power" (213).

Just over a year after my *New York Times* op-ed was published, the United States was experiencing the full force of the global COVID-19 pandemic, necessitating shelter-in-place and stay-at-home orders across the country. Seemingly overnight, tens of millions of Americans were told to work, study, shop, play, and communicate from home and online. The confident move to entirely digital operations, however, presumed everyone had access to high-performance broadband. For those who did not, so many of whom live and work in rural America, COVID-19 was another painful reminder of the digital divide and rural penalty. The pandemic has reinforced the conclusions of this book and amplified the stakes to near life and death levels. A headline in the UK-based *Guardian* newspaper underscored this amplification: "US Digital Divide 'Is Going to Kill People' as COVID-19 Exposes Inequalities" (Holpuch 2020). It has never been more crucial to connect all Americans (and indeed, the world) with low-cost, high-performance broadband. Outlining this vision is the point of this conclusion to the book. I begin with a description of rural broadband in both a COVID and post-COVID world to set the stage and remind us of the consequences of ineffective broadband policy. For many, the pandemic has cast a spotlight on the digital divide for the first time and spurred calls to action across the political spectrum. Unfortunately, it has also demonstrated the ineffectiveness of the FCC, a lack of urgency by Congress, and how deeply entrenched this country is with the political economy of Big Telco. Indeed, the politics of "good enough" have never been more apparent. To exemplify this last point, I then move to events on the horizon—5G and LEO satellites—and cut through the hype to discuss the applicability of these technologies for rural communities. Last, I connect the dots by offering my solution to the rural-urban digital divide in the United States: a renewed national rural broadband plan.

Rural Connectivity in a (Post-)COVID World

A Stay-at-Home Nation
The first full draft of *Farm Fresh Broadband* was completed a month before stay-at-home and shelter-in-place mandates were announced for most of the United States in light of the global spread of COVID-19. In the months to follow, the pandemic forcefully and painfully demonstrated to the world the critical importance of broadband and highlighted the gaps in broadband deployment and failures of broadband policy. Moreover, it reminded

us that the digital divide is not only a rural issue, or an American issue, but an issue caused by and exacerbating a host of socioeconomic inequalities.

In the United States, the earliest concerns for broadband during the pandemic focused on the capacity of networks to handle a country working and studying from home (see, e.g., Alba and Kang 2020). Debates centered particularly on DSL and cable, with critics suggesting that DSL was not up to the task of a country on Zoom (Reardon 2020) and predicting that cable would suffer from intense network congestion (Heaven 2020). This line of criticism was eventually replaced with concern for those without broadband. The National Bureau of Economic Research, for instance, reported that "the combination of having both high income and high-speed internet appears to be the biggest driver of propensity to stay at home" (Chiou and Tucker 2020). The study concluded with a damning statement on the digital divide: "The digital divide . . . appears to explain much inequality we observe in people's ability to self-isolate" (Chiou and Tucker 2020). An article from the Associated Press summed the issue up best: "Those without broadband struggle in a stay-at-home nation" (Arbel and Casey 2020)

The infrastructure gaps in rural America, so familiar to us in this book, were highlighted in dozens of news articles and reports (e.g., Alleven 2020a; Higgins 2020; Kang 2020; Kaur 2020), as were the cost and subscription gaps across the country (e.g., Chao and Park 2020; Seifer and Callahan 2020). In light of the shocking rise in unemployment caused by the pandemic, calls were made to expand the Lifeline program (Sohn 2020), as were stirring arguments that broadband be regarded as a civil right. Connecting broadband access with the Black Lives Matter and social justice movements that gained such prominence and presence in the spring and summer of 2020, Reverend Al Sharpton and FCC commissioner Geoffrey Starks, Vanita Gupta, Marc Morial, and Maurita Coley wrote in *Essence*:

> The Internet is also a powerful tool for building movements—from allowing our communities to share our narratives in new ways to amplifying mobilizing efforts. Online activism is a critical part of how we shared stories about the lives of George Floyd, Ahmaud Arbery, Breonna Taylor, Trayvon Martin, Sandra Bland, Michael Brown and so many more. Connectivity helps bend the arc of history toward justice. (Sharpton et al. 2020)

Without a doubt, unconnected and underconnected students received the bulk of commentary and press coverage related to COVID and the digital divide (see, e.g., Fishbane and Tomer 2020). References to the homework gap, discussed the introduction to this book, resurfaced, as did stories of parents driving their children to the parking lots of McDonald's to use the Wi-Fi (Fishbane and Tomer 2020; Kang 2020). Worse were the stories of students unable to attend classes or complete homework because of a lack of broadband (Khazan 2020). Again, we were reminded that the homework gap is particularly acute for low-income and minority students and students living in rural communities (Auxier and Anderson 2020). In the remote Canadian north, with a large percentage of Indigenous peoples, for instance, a lack of broadband access meant that students were communicating with teachers via landline telephones and fax machines (Jody Porter 2020).

Amid these concerns, however, we also saw greatness and examples of communities and companies doing what they could to connect the unconnected. Examples include Wi-Fi enabled school busses parked in unconnected neighborhoods (Mattise 2020) and the construction of solar-powered public hotspots (Higgins 2020). An innovative public-private partnership in Champagne, Illinois, brought fixed wireless to a mobile home park (Marcattilio-McCracken 2020) while a wireless mesh network connected a neighborhood that included over 2,000 students in San Rafael, California (Quaintance 2020). Across the country, libraries extended their Wi-Fi signals into parking lots and alongside schools began requesting and loaning out so many hotspots that the country experienced a national shortage of the devices (McGill 2020; Opalka et al. 2020). In the corporate space, dairy cooperative Land O'Lakes (2020) launched the American Connection Project Broadband Coalition, a group of over forty companies, including many familiar to readers of this book such as the NTCA, NRECA, and NFU, to "advocate for public and private sector investment" in broadband.

As the agency with the assumed authority to correct the digital divide, the FCC was called upon to do just that. Unfortunately, the Commission has surprisingly little power over broadband since it effectively regulated itself out of regulating ISPs when it reclassified broadband as a Title 1 Information Service (FCC 2018c; Sohn 2018). To its credit, the FCC relaxed conditions on the E-Rate program, allowing recipients of E-Rate (i.e., schools and libraries) to expand their Wi-Fi signal reach (FCC 2020h). Without direct authority over providers, however, the FCC's hands were tied to do much more. As a result, the FCC asked (but could not require) providers to promise not to disconnect customers who could not afford to pay for their broadband subscriptions. This became the "Keep Americans Connected" pledge (FCC 2020c). In addition, some of the largest ISPs, like Charter and Comcast, offered free or discounted connectivity to low-income households and families with students (Castaneda 2020), while others mobilized portable cell towers to increase connectivity (Fletcher 2020). Acknowledging the rise in network traffic, many ISPs also lifted data caps (Hachman 2020). As the pandemic raged into the summer of 2020, however, the Keep America Connected pledge expired, and many ISPs reverted to charging full price for subscriptions and reinstituted data caps (Hachman 2020). The FCC is currently powerless to stop this from happening.

For all of its horrible outcomes, and despite FCC impotence, the COVID-19 pandemic has created an opening for impactful congressional action into broadband deployment and access. Without doubt 2020 saw a critical moment for broadband, and closing the digital divide gained new prominence among leading lawmakers (McKinnon and Tracy 2020). Vexingly, in fall 2020 when this chapter was revised, the life-and-death need for universal broadband and the community commitment to connectivity that has been demonstrated throughout the pandemic has not translated into meaningful policy or legislative action at the federal level.[2]

The 2020 Coronavirus Aid, Relief and Economic Security Act (CARES Act) represented a decisive opportunity for Congress to quite literally put its money where its mouth is when it came to broadband. In 2017, the FCC had published a report estimating that it would cost $80 billion to connect the unconnected with fiber in the country (de Sa 2017); at that time, $80 billion seemed like a lot of money. Indeed, it *is* a lot of money. Then the COVID-19 pandemic emerged and Congress began debating a multitrillion-dollar stimulus package. Suddenly, the optics changed, and $80 billion was peanuts when placed against $2 trillion. This was Congress's chance to invest in broadband. When it was passed, however, the CARES Act had failed to harness the moment. The stimulus package contained only minimal allocations for broadband: $200 million for the FCC's rural health care program, $100 million for the ReConnect Program (see chapter 3), $50 million for the Institute of Museum and Library Services, and $25 million for the RUS's Distance Learning Telemedicine and Rural Broadband Programs. These allocations total $375 million, or just 0.01875 percent of the $2 trillion package.

Despite the omission of substantive broadband subsidies, twenty-four states used general CARES Act monies to fund broadband deployment and subsidize broadband subscriptions (Philbrick 2020). Together these states spent upward of $1 billion in CARES Act funding on broadband access, according to a report by Maine's Economic Recovery Committee (Philbrick 2020). Wyoming, for instance, "spent the greatest proportion of its CARES Act funding on broadband expansion, at more than 8 percent," while "as a nominal figure, Alabama dedicated the largest appropriate of any state to broadband" at $103.4 million (Philbrick 2020, 2–3). Much of the funds spent by these twenty-four states went to rural broadband deployment, including both FTTH and fixed wireless projects (Philbrick 2020).

More aggressive broadband funding packages were proposed in the Health and Economic Recovery Omnibus Emergency Solutions Act (HEROES Act) in May 2020 and the Accessible, Affordable Internet for All Act passed two months later. The former allocated $7.5 billion for broadband deployment and access. This included a $1.5 billion injection into the E-Rate program for connectivity and devices, $4 billion to create an Emergency Broadband Connectivity fund that would provide up to $50 per month toward a broadband subscription for those laid off or furloughed because of the pandemic, and $2 billion to the FCC's Rural Health Care Program. It would also prohibit telecommunication providers from discontinuing service to those whose income was impacted by COVID-19 (thus codifying the FCC's Keep America Connected pledge) (Akin Gump Strauss Hauer & Feld 2020). Lastly, the HEROES Act would allocate $24 million to the FCC so it could begin revising its broadband mapping data collection methodology as was ordered in the 2020 Broadband DATA Act.

The Accessible, Affordable Internet for All Act went even further. Introduced in June 2020 by congressperson Jim Clyburn, who also founded and heads the Whip's Rural Broadband Task Force of House Democrats, the act proposed $100 billion for broadband deployment and access. This included $80 billion for broadband deployment in unserved areas and "low tier" areas ("low tier" defined as areas with speeds between 25/25 and 100/100mbps), $5 billion for broadband loans administrated by the NTIA, and $1 billion for state grant programs. The remaining funds included $9 billion for Lifeline (to offer $50 a month to subsidize broadband subscriptions for low-income households and $75 for those on tribal lands) and $5 billion for E-Rate. Unfortunately, neither of these bills has become law at the time of writing.

Both the inadequate broadband provisions in the CARES Act and the failure to pass the HEROES and Accessible, Affordable Internet for All Acts represent missed opportunities in this critical broadband moment. As a result, and as the pandemic rages on, millions of Americans—students, rural communities, low-income households, among them—continue to bear the burden of the digital divide.

Waiting for 5G

But wait! Isn't 5G just around the corner? Won't this new technology save rural America from the grips of the rural penalty? This, unfortunately, is a narrative I have heard all too often in rural communities. It is a narrative hammered home by the largest mobile providers eager to secure regulatory favor and subsidy. Take T-Mobile and Sprint. Rural 5G deployment was the cornerstone of the promises made by these companies in exchange for permission from the FCC and Department of Justice to merge. "The promises [of 5G] are endless," said T-Mobile CEO John Leger in 2019: "If our merger is approved, we will have a network that has the scale necessary to challenge the status quo and help bridge this divide for those Americans who have been left to look across it." The pair promised to cover 95.8 percent of the country's 62 million rural residents with its combined 5G network—if, and only if, they are allowed to merge (Legere 2019). Their wish was granted late in 2019, and the company now operates as the "New T-Mobile."

The problem with this promise is that the New T-Mobile's 5G network is based on low-band spectrum that will only offer customers an enhanced 4G LTE experience (Brodkin 2019). As discussed in the introduction to this book, in urban America, AT&T and Verizon are rolling out high-band or millimeter wave 5G networks. These connections can support the IoT so lauded by the big tech industries, from Apple and Google to Deere and Tesla—everything from driverless cars to toasters and Fitbits. While the possibilities are promising, they are only viable for densely populated and wealthy urban centers. Not only will high-band 5G networks require an expensive new phone, but millimeter waves can only travel 800–1500 feet before requiring a digital boost in the form of a "small cell": a pizza-box-size signal repeater. Small-cell deployment has already received the wrath of suburban households unwilling to cede their light posts and utility poles to more technology (recalling the acronym NIMBY, "not in my backyard"). They also require a full fiber backhaul, which comes with its own expense (Crawford 2019). Their limited geographic reach and extreme cost (approximately \$3,000 per cell[3]), moreover, make them unfeasible for rural deployment.

Despite the geographic limitations of millimeter waves and small cells, the industry hype over 5G is real and aggressive, so real in fact that in 2018 the FCC relaxed the power of municipalities over small-cell deployment (FCC 2018b) and in 2019 created a $9 billion fund for rural 5G deployment (FCC 2020e). To be sure, 5G network deployment is happening, but it is happening at a glacial pace even in the densest and wealthiest centers (Chiaraviglio et al. 2016). Like electricity, telephony, and earlier broadband technologies, this deployment asymmetry threatens to keep rural America off the grid, either because rural counties may decide to hold off their broadband plans to wait for 5G, or because it will be seen as an unwise financial investment by mobile carriers.

If not in the ether, perhaps the solution to the rural-urban digital divide will be found in the heavens? LEOs, made famous by Elon Musk's SpaceX, have garnered much attention. LEOs are satellites that sit closer to earth (99 to 120 miles) compared to the geosynchronous satellites currently employed (22,000 miles) (Cooper 2019). They can be deployed in the thousands and are touted as a novel fix to the digital divide (Cooper 2019). Presently, SpaceX has permission from the FCC to launch as many as 12,000 satellites into orbit as part of its Starlink project, while Amazon, through its Project Kuiper, has plans for a 3,236-satellite network (Jon Porter 2019). Thanks to some hefty negotiations, LEO providers are also eligible to compete for funding from the RDOF, which began accepting bids in October 2020. The benefit of LEOs is both their proximity to Earth, which minimizes latency compared to geosynchronous satellites, and their volume, which can in theory cover the entire country through a constellation pattern. Early trials of Starlink saw speeds upward of 100 Mbps download with low latency, a victory according to SpaceX, but substantially slower than what was initially promised (Grush 2020).

To be sure, both 5G and LEOs offer important alternatives to fixed broadband, but they are alternatives with uncertain deployment for rural America. The concern is that rural communities may end up in a *Waiting for Godot* situation if they hold off their broadband plans in favor of these future networks.[4] The technology already exists today to connect the unconnected; the problem is getting it to those who need it most.

The overarching issue facing the rural-urban digital divide is not about technology, nor is it about money. It's about policy and politics, or rather the lack of policy and the abundance of politics. It's the politics of incumbency that allowed CenturyLink and Frontier to garner millions of dollars a year in CAF II subsidies and then fail to live up to their commitments (Brodkin 2020c; Taglang 2019). It's the politics of technological neutrality that allowed ViaSat to come out as one of the largest winners in the CAF II auction, despite offering connections that fail to live up to the definition of broadband. It's the politics of power that has stymied attempts to revise broadband mapping and replace Form 477 whereby ISPs exaggerate rural connections. It's the politics of complacency that has created a bifurcated policy apparatus between the FCC and USDA that results in inefficiencies and knowledge silos. It's the politics of economics that sees the largest telecommunications companies shift their sights to 5G and abandon their fixed networks and rural customers. It's the politics of "good enough" that keeps the status quo in place and withhold alternatives. These politics persist because of a lack of policy, enforcement, and oversight. To put an end to what I call the "politics of the status quo" and to efficiently and democratically distribute the billions of dollars that have been earmarked for rural broadband deployment in the next decade, the United States needs a new national rural broadband plan.

A National Rural Broadband Plan

In chapters 1 and 3, I recounted the history of rural electricity and telephony and focused on the vision of the 1936 Rural Electrification Act and the REA. I discussed the success of these programs that brought electricity and later telephony to rural America. We need to reclaim this energy and focus for rural broadband deployment. To do so requires a plan of action as much as it requires the investment of resources. In fact, the latter is all but useless without the former.

As I intimated in chapter 3, it should be the USDA (rather than the FCC) that authors America's rural broadband plan. The FCC, by its structure and organization, has proven itself incapable of rising above the partisan politics that hold the United States in a vise-like grip in 2020.[5] The politics of the FCC sway with the politics of the president, and the FCC of Ajit Pai has demonstrated that it will simply sweep aside the decisions of previous administrations with which it disagrees. The 2018 repeal of network neutrality and the heightened rate of media consolidation are two excellent examples of these politics.

There is strong historical precedent for my argument that the USDA create America's new rural broadband plan. Recall that it was the USDA, once it merged with the REA, that succeeded at electrifying the countryside in the 1940s. So successful was this program that the USDA was charged with connecting rural America to the telephone grid in the 1950s. This program was also lauded as a tremendous success. In both of these moments, the USDA championed local cooperatives to assume the challenge of connection. Recall that the department was tasked with coauthoring the country's original rural broadband plan in 2008. Under President Barack Obama, the USDA also cochaired the 2015 Broadband Opportunity Council and was tasked with awarding billions of dollars in stimulus funding in 2009 for broadband deployment. More recently, Congress awarded the USDA $600 million to continue the ReConnect program. This is on top of the $800 million in broadband loans and grants the department annually administers. Last, the USDA has been a vocal champion of rural broadband for precision agriculture and has worked to ensure that farms are given priority in awarding ReConnect grants and loans.

True, the USDA has not been successful at all of its broadband initiatives. As chapters 2 and 3 demonstrated, the FCC ended up authoring the 2009 rural broadband strategy, which was eventually eclipsed by the FCC's 2010 NBP. The USDA was also accused of mismanaging the funds it received from the 2009 ARRA. As chapters 3 and 4 demonstrated, small ISPs also find it exceedingly difficult to apply for USDA loans and grants, and, if successful, have difficulties managing the bureaucratic expectations. As chapter 5 recounted, the USDA and RUS have bought into the rhetoric of rural broadband for precision agriculture without considering the role of, and unequal benefits to, industrial agriculture and Big Telco. The department is also subject to its own political squabbles, including a recent controversial decision to relocate the USDA's Economic Research Service and the National Institute on Food and Agriculture, and their combined 600 employees, from Washington, DC, to Kansas City, Missouri. Critics suggest that the move is politically motivated because the findings of these two research units tend to disagree with the administration's political agenda (Morris 2019).

What the USDA does bring to the table is a connection with rural America not consistently found at the FCC. The FCC does not have an office of rural broadband,[6] nor has its support of rural connectivity been dependable. As already noted, the FCC's own structural ties to incumbent politics make charging it with such an important task unadvisable at best. In contrast, the USDA has field offices in every state and many territories and is mandated by law to champion rural America. As the RUS (1996) wrote to the FCC,

> RUS is in a unique position to comment on rural America's telecommunications needs. The Agency's goal has always been to provide every rural household with affordable service. Our point of reference is the urban and suburban subscriber. We have sought to ensure that RUS borrowers provide telecommunications service that works like, sounds like, and costs like the urban and suburban customers' service. Since this is much harder to do in low density areas, RUS has created its own practices and standards which addressed the rural challenges. (2)

My proposal for a more purposeful role for USDA in setting rural broadband policy finds support with a number of rural broadband stakeholders. The AFBF, Texas Statewide Telephone Cooperative, and NTCA have in the past supported an enhanced policy role for the USDA (see chapters 2 and 3). In 2017, Rep. Jared Huffman (D-CA) introduced the New Deal Rural Broadband Act. This act would establish a rural broadband office within the USDA and create the position of under-secretary for rural broadband with responsibilities to

- administer all rural broadband-related grant and loan programs currently administered by the Rural Utilities Service,
- conduct specified outreach and coordination activities, and
- conduct and release to the public an inventory of federal and state property on which a broadband facility could be constructed. (New Deal Rural Broadband Act of 2017, first summary; see also Ali 2019a)

Among those who agree with the active presence of the USDA in broadband policy, however, the strongest endorsement comes from Eric Frederick of Connected Nation, who simply stated in an interview with me: "They know rural development" (see chapter 3). History certainly agrees with Frederick on this point (see chapter 1).

The USDA should be charged by Congress to author a new rural broadband plan, just as Congress charged the department with drafting the 2009 rural broadband strategy and cochairing the 2015 Broadband Opportunity Council. This time, however, the USDA must be given the necessary resources to complete the task.

The country's rural broadband plan and ensuing rural broadband policy framework should take into account the following recommendations:

- Create an office of rural broadband at the USDA and within the RUS and empower the office to be the point agency for rural broadband policy.
- Coordinate with the FCC and NTIA to harmonize broadband goals and funding priorities. This would include eliminating the cannibalism that resulted when the RDOF program excluded ReConnect and state-funded areas from eligibility (see chapter 2).
- Manage congressional appropriations for rural broadband deployment, such those proposed in the Accessible, Affordable Internet for All Act (with the exception of funds earmarked for the FCC's designated programs and those allocated for specific purposes of the NTIA).
- Set an ambitious definition of broadband that maintains the ethos of technological neutrality but sets forward-looking speed and latency thresholds, such as the 100/100 Mbps suggested by Jonathan Sallet (2019b). This way technologies are not excluded prima facie but must be able to deliver these speeds on day one.
- Revise the broadband mapping methodology and stop relying on industry self-reports. Instead, the new methodology would amalgamate a host of different data sets, including the crowdsourcing that M-Lab has found so valuable.
- Include a challenge process so rural communities have an avenue to protest if they are overcounted in broadband maps.
- Work with states to ensure every state has a fully staffed broadband office. Federal funds allocated to states for broadband should only be released once an office has been established, since the presence of state offices has been proven to have some positive impact on deployment (Whitacre and Gallardo 2020).
- Eliminate state barriers to cooperative and municipal broadband that have, by and large, been passed at the behest of Big Telco.
- Democratize the funding process and stop privileging the largest telecommunications companies. This includes allowing cooperatives, municipalities, and new entrants to bid on funding, and to draft strong punishments for defaults.
- Streamline and simplify the funding process.
- Fund digital inclusion efforts, specifically meaningful subsidies for low-income households, computer access and digital skill development. Broadband connections that sit unused and unaffordable are all but worthless.

- Reconsider the nation's 5G strategy and encourage mid-band 5G deployment in rural America.

With this list of recommendations, I am not suggesting that the USDA be the agency that allocates all $10 billion in broadband subsidies. To the contrary, the FCC and NTIA have vital roles to play in broadband funding and policy. The USF, for instance, has been crucial in providing the operating capital for "high cost" networks.[7] Instead, I suggest that the USDA become the lead agency to set rural broadband policy and coordinate between the federal and state agencies invested in rural broadband deployment and adoption. COVID-19 demonstrated that the key to universal broadband is not to be found with one company, one fund, or one program. Instead, the rural-urban digital divide will only be corrected through a multistakeholder, "all-hands-on-deck" approach: cooperatives, municipalities, small ISPs, and even the large carriers; the USDA, FCC, NTIA, and state broadband offices; and, of course, local communities. This recalls Italian economist Antonio de Viti de Marco's argument in his famous 1936 book *First Principles of Public Finance*, that the ideal provision of public goods (or club goods, given our discussion in chapter 1) is not the exclusive domain of either the market or the state, but rather one of cooperation (see also Medema 2009). In the case of rural broadband deployment, the country needs a multistakeholder broadband system guided by insightful public policy that recognizes the value of public networks, private networks, and public-private partnerships. It would be a network of networks funded through diverse streams, including public funding at the federal, state, and local levels, industry cross-subsidy through the USF, and private market-based investment.

It is only by recognizing the multitude of possibilities, rather than assuming a market-based approach a priori, that the rural-urban digital divide will finally be resolved. America achieved near-universal service with electricity and telephony because of a multistakeholder approach, led by smart and deliberate federal policy. We've done it before. We can do it again.

A Global Divide

This book arrives at crucial juncture in the history of the rural-urban digital divide. Broadband is having a moment in contemporary American consciousness. It was part of five Democratic presidential candidates' policy platforms in the summer of 2019 (Joe Biden, Pete Buttigieg, Amy Klobuchar,

Bernie Sanders, and Elizabeth Warren) and even made it into the 2020 State of the Union and 2021 proposed federal budget. With the onset of the COVID-19 pandemic, affordable, high-performance broadband has never been more important. It is now near-universally acknowledged that access to broadband and an accompanying digital skill set is not a luxury or a novelty but a necessity. As Jan van Dijk observed in his 2020 book, *The Digital Divide*, "Digital media are used for every act, purpose or need in society. Increasingly, access to and use of digital media is needed to participate as a worker, entrepreneur, student, consumer or citizen, or in any other role in contemporary society" (5). So important is access to, and accessibility of, digital media that internet access has been framed as a civil right (Sharpton et al. 2020) and a human right (Jasmontaite and de Hart 2019; Peacock 2019; Reglitz 2019). The United Nations (2016) proclaimed internet access a human right in 2016, and the EU (2019) has set a goal of connecting every village and city with free wireless internet by 2020 (see also C. Howell and West 2016). In 2010, Finland became the first country to make broadband a legal right (*BBC* 2010). Canada, meanwhile, has proclaimed broadband a "basic telecommunications service," while in Germany broadband is a "basic right" (Canadian Radio-Television and Telecommunications Commission 2016; Wunsch 2013). This reminds us that the challenges of rural broadband deployment are not unique to the United States; it is a global issue. It also demonstrates, however, how far the United States has to go in framing broadband as a right rather than a commodity.

Paradoxically, this book also comes at a time in which global broadband penetration rates have slowed dramatically (ITU 2015). Here in the United States, the rate of fixed broadband adoption has slowed (Broadband Tech Report 2019), and the FCC (2020a, 2019a) has congratulated itself for narrowing the digital divide. But the digital divide is not shrinking, nor may it ever. Jan van Dijk (2020) argues that even when problems of access and accessibility are resolved, "the digital divide is here to stay" (4). This is because new technologies, like 5G and LEOs, will be deployed first in urban, wealthy areas, with rural communities again an afterthought. Even without these new technologies, urban speeds continue to rise while rural speeds remain stagnant (with the many exceptions noted throughout this book). Global internet connectivity hovers just above 50 percent (ITU 2019), with dramatic differences between the urban and rural. A report by the Alliance for Affordable Internet (2020) found that only 14 percent of rural areas in the Global South have internet access. In this new COVID era, broadband has proven itself a critical utility and lifeline, but one that can only be universally realized with serious investments in policy, planning, and resources. Without this investment both in the United States and throughout the world, the gap between rural and urban broadband is set to fulfill van Dijk's pronouncement.

Better Broadband

I am continually struck by something Bernadine Joselyn, director of public policy and engagement at Minnesota's Blandin Foundation, told me in her interview for this book: "Everything is better with better broadband." I repeat it here and throughout this book to shed a clear, bright light on a complex issue. This is not a technologically determinist argument—the mere existence of broadband in a rural community does not correct rural inequity ipso facto, although the lack of connectivity certainly does not help. Instead, I invoke this phrase to demonstrate how broadband, rural broadband, and broadband policy are lived and experienced (or not) on the ground in rural communities throughout the country. Broadband policy (such as it is) may be written in Washington, with legal prose refined at the headquarters of AT&T, Verizon, CenturyLink, Windstream, and Frontier, but it is lived in Luverne, Minnesota; McKee, Kentucky; and Mineral, Virginia. At its farthest reach, Joselyn's phrase parallels conversations occurring at the United Nations and the European Union about whether internet access can be called a human right, akin to the rights to information, education, and participation. Closer to home, it reminds us that at the end of the day, broadband is not about policy, politics, technology, or money; it is about people. When deployed democratically and harnessed inclusively, everything can be better with better broadband, from homework to work to voting to health to talking with grandma. "Everything is better with better broadband" is a call to end the rural-urban digital divide and the multitude of digital divides brought on by inequality in the United States.

Notes

Introduction

1. The Australian NBN began as an ambitious plan to connect the entire country to a publicly funded fiber-optic network that would operate as a wholesaler. According to B. Howell and Potgieter (2020), "It represented the largest infrastructure investment ever undertaken by the Australian government" (2). The NBN was proposed in 2007 and is scheduled for completion in 2021. In the intervening years, costs have skyrocketed (from an original estimate of A$15 billion [$11 billion] to a final price tag of A$49 billion [$35 billion]) and the initial full fiber plan morphed into a "mixed technology model" in 2013, much to the chagrin of many in the country (Marashallsea 2017).

2. The exact amount is difficult to parse. The FCC, through the USF, subsidizes broadband across the country to the amount of $8.3 billion. At least $5 billion is exclusively earmarked for rural broadband deployment, and the remaining $3.3 billion is spread among low-income consumers, schools, libraries, and health centers (including rural). Adding the $800 million in grants and loans from the USDA/RUS, and the recent $600 million ReConnect Loan and Grant Program, I estimate that *at least* $6 billion is spent specifically on rural broadband per year.

3. I have authored several of these papers and op-eds (Ali 2018, 2019a, 2019b, 2019c; Ali and Contractor 2020).

4. See, e.g., Kang (2017), Levitz and Bauerlein (2017), Strover (2018), Whitacre (2016), and everything by Jon Brodkin, Jason Koebler, and Karl Bode in the References, to name just a few of my favorite journalists covering rural broadband and broadband policy.

5. Jonathan Sallet, senior fellow at the Benton Institute for Broadband and Society and former FCC general counsel, argues in his pathbreaking report *Broadband for America's Future* (Sallet 2019b) that not all broadband is "high-performance." He defines high-performance broadband as "fixed broadband networks are fit for the future; they provide fast, symmetrical upload and download speeds, low latency (moving data without noticeable delay), ample monthly usage capacity, and security

from cyberattacks. Networks that, once installed, can be easily upgraded as the demand for greater broadband performance continues to increase" (12).

6. Hindman (2000) offers a different origin story. It is not my intention to do a full review of the literature on the history of the digital divide. For early foundational research, see Compaine (2001), Norris (2001), and J. van Dijk (2005). For later research see J. van Dijk (2020).

7. There is still dial-up internet, where connectivity is achieved through regular copper phone lines. This is how we accessed the internet in the 1990s (think AOL or CompuServe). Dial-up does not meet the definition of broadband, which is why it is not included in the discussion beyond this footnote. Still, CNN reported in 2015 that 2.1 million Americans subscribed to AOL's dial-up service (Pagliery 2015).

8. Some readers may be curious about the difference between DSL (digital subscriber line) and ADSL (asymmetric digital subscriber line). ADSL is a prevalent type of DSL that delivers substantially more bandwidth downstream than upstream (hence, asymmetric) (Newton 2018). This makes it useful for website browsing. Both DSL and ADSL, however, are substantially slower than cable or fiber, and networks using DSL have been criticized for being unable to handle the added internet traffic that came with the COVID-19 pandemic (Bray 2020). For shorthand, I refer to both DSL and ADSL as "DSL."

9. Fixed wireless can also make use of microwave backhaul rather than DSL or fiber.

10. There has unfortunately not been much development on the Airband Initiative since its announcement in 2017.

11. The G in 3G/4G/5G stands for "generation," with the number referring to the generation of cellular connectivity. LTE stands for "long-term evolution." Each iteration means faster and greater capability. For example, 4G's fastest speeds hover around 120 Mbps (although most are around 20 Mbps) while 5G can reach gigabit speeds (Hu 2018).

12. Here, I am discussing internet service provided by companies operating geosynchronous satellites. This is different from the newly launched low-Earth orbital (LEO) satellite networks owned and operated by providers like Starlink. I address LEO satellite provision in the conclusion.

13. In 2020, Australian provider Telstra teamed up with Ericsson to extend the signal range of an LTE tower from the standard 100 kilometers (60 miles) to 200 kilometers (120 miles) (Duckett 2020).

14. This is slowly beginning to change with the FCC's summer 2020 auction of 70 MHz in the 3.5 GHz Citizens Broadband Radio Service band (Alleven 2020b).

15. "Telehealth is defined as the use of electronic information and telecommunication technologies to support long-distance clinical health care, patient and professional

health-related education, public health and health administration" (Mississippi Media Center qtd. in Gallardo 2016, chap. 3, sec. 4).

16. The US Census Bureau uses a negation definition: Everything that is not urban is rural. The USDA is slightly different, and its definition accounts for 72 percent of the country's landmass and 46 million residents.

17. See Mosco (2009), Sarikaskis and Shade (2008), Mansell (2012), McChesney (1995), Pickard (2019), and Winseck and Jin (2011). For more on the foundations and foundational figures of the political economy of communication, see Wasko, Murdock, and Sousa (2011).

18. It is impossible in the space allotted to give a full account of the scholarship along this axis. In addition to those cited here, see Castells (2013), Mansell (2004), and the edited collections of George (2017) and Wasko, Murdock, and Sousa (2011).

Chapter 1

1. In a contemporary analog, as part of his modern Green New Deal proposal, 2020 presidential candidate Bernie Sanders included $150 billion for broadband deployment (see Berman 2019).

2. For rural electrification, see Kline (2000 [also on rural telephony]), Tobey (1996), Brown (1980), Rudolph and Ridley (1986), and Nye (1990). For rural telephony see, MacDougall (2014), Fischer (1992), de Sola Pool (1977), Wu (2010), and John (2015). I also relied on a number of primary documents provided by both the NRECA and NTCA. I am grateful for their support of this research.

3. See Copeland and Severn (1985) for more on rates and elasticity.

4. In most of these conversations, "rural" was taken as synonymous with "farm" and "farming," since farmers (presumably living in rural America) constituted a full 35 percent of the population in 1910 (Kline 2000).

5. Tobey (1996) makes the important point that the housing authorities and Public Works Administration created by the New Deal also played tremendous roles in rural electrification, especially before the REA in 1935.

6. Interestingly, Lyndon B. Johnson was originally offered the directorship of the newly recast REA in 1939.

7. For a terrific history of the concept of universal service in telecommunications, see Mueller (1997).

8. Bell has long been associated with the image of an octopus, a metaphor created by Paul Latzke in his 1906 history of the telephone *A Fight with an Octopus*. An octopus is an apt metaphor because Bell controlled so much of the networked architecture of telephony (Kline 2000).

9. Dual service gave rise to Bell's push for "universal service," which at the turn of the century referred not to the modern iteration of a phone in every home, but rather a singular (universal) network (MacDougall 2014; Mueller 1997).

10. As MacDougall (2014) writes of the Kingsbury Commitment,

> This is, in microcosm, how competition in American telephony came to an end. Support for dual service dwindled among business users and municipal governments, the very groups that had given birth to independent telephony. State utility commissions pressed for interconnection, on the grounds that regulation could do the work competition had once done. City by city and market by market, swaps and consolidations . . . eliminated direct competition, parceling out the country to Bell and independent firms. (217)

11. This is not without debate. Many argue that broadband and internet access is a public good (Pickard and Berman 2019; Tarnoff 2016; Treacy 2017)

12. Other excludable traits of broadband include redlining, deployment prohibitions, distributed denial of service attacks (DDoS) and internet shutdowns (see Raymond and Smith 2014, 21).

Chapter 2

1. This chapter is based on a previously published article in The International Journal of Communication (Ali 2020)

2. At his most melodramatic, Marx compared capitalism to a vampire: "Capital is dead labour, that, vampire-like, only lives by sucking living labour, and lives the more, the more labour it sucks" (Marx [1867] 2004, chap.10, sec. 1).

3. The FCC's 2018 removal of network neutrality protections and the reclassification of internet service providers under Title I of the 1996 Telecommunications Act has potentially jeopardized the FCC's regulatory power to enforce universal service (Brodkin 2017c).

4. There are, of course, other salient sections of the act, including Title II, which governs common carriers, and Section 303, which governs the relationship between common carriers and overbuilders, but Sections 254 and 706 are the most relevant to this book.

5. The FCC (2006) did release a report titled *Lands of Opportunity: Bringing Telecommunications Services to Rural Communities*, but it focused only on the Appalachian region, the Delta region, and Alaska Native villages.

6. I define "Big Telco" as the collection of the largest national telecommunications companies: AT&T, T-Mobile, Sprint, Verizon, CenturyLink, Windstream, and Frontier (see Gallardo and Whitacre 2019; C. Mitchell and Trostle 2018).

7. To be fair, telecommunication companies are also investing in fiber deployment.

8. The NTCA was formerly called the National Telephone Cooperative Association and (briefly) the National Telecommunications Cooperative Association.

9. As Teun van Dijk (1993) notes of discursive formations,

> Here we touch upon the core of critical discourse analysis: that is, a detailed description, explanation and critique of the ways dominant discourses (indirectly) influence such socially shared knowledge, attitudes and ideologies, namely through their role in the manufacture of concrete models. More specifically, we need to know how specific discourse structures determine specific mental processes or facilitate the formation of specific social representations. Thus, it may be the case that specific rhetorical figures, such as hyperboles or metaphors, preferentially affect the organization of models or the formation of opinions embodied in such models. (258–259)

10. In 2014, the FCC unveiled a series of grants for rural broadband experiments as part of the Connect America transition, using Connect America funds. The FCC allocated $100 million for these experiments that would be located in areas occupied but unserved by the major telecommunications companies. The "experiment" here was "to test on a limited scale the use of a competitive bidding process to award support to provide robust broadband to serve fixed locations using both wireline and wireless technologies" (FCC 2014c, 5). This is the same process found in the CAF II auction.

11. The CAF I included allocations of more than $600 million to price-cap carriers to provide service of 4/1 to unserved areas.

12. Previously called the *Broadband Progress Report*.

13. The NTIA was ordered to create a National Broadband Map in the 2009 American Recovery and Reinvestment Act. It managed the map until 2014. The map was dormant until revived by the FCC in 2017.

14. It should be noted that RUS funding (as detailed in chapter 3) does not use the National Broadband Map but rather asks that applicants draw their own maps to detail unserved and underserved areas—a task that comes with its own burdens and expenses.

15. As this book was in the final production stages, Congress passed the 2021 Consolidated Appropriations Act, Section 906 of which earmarks $65 million for the FCC, specifically for broadband mapping.

16. Feld is referring to House speaker Nancy Pelosi's exaggerated, condescending applause during President Trump's 2019 State of the Union Address (see Chiu 2019).

Chapter 3

1. This chapter is based on a previously published article in *Telecommunications Policy* (Ali and Duemmel 2018).

2. This of course would be dwarfed by the $2 trillion stimulus package from the CARES Act that was necessitated by the spread of COVID-19 in 2020.

3. The 2021 Consolidated Appropriations Act, a response to the COVID-19 pandemic, came close, allocating $7 billion for broadband access and deployment.

4. In 1994 Congress passed the Federal Crop Insurance Reform and Department of Agriculture Reorganization Act. As part of this act, the responsibilities of the REA were transferred to the newly created RUS, whose mandate also included water (both potable and sewage) in addition to telecommunications and electricity.

5. The Farm Bill is an omnibus bill that governs and funds the federal government's many agriculture and foods programs. It is renewed every four to six years. The most recent Farm Bills were passed in 2018, 2014, and 2008 (R. Johnson and Monk 2017).

6. HR 2330: Agriculture, Rural Development, Food and Drug Administration, and Related Agencies Appropriations Act of 2002.

7. Formally known as the 2002 Farm Security and Rural Investment Act.

8. The exception here is the ReConnect Program whose parameters were set by Congress in the 2018 Consolidated Appropriations Act.

9. Section 779 of the 2018 Consolidated Appropriations Act expressly forbids the funding of overbuilders.

10. I obtained this document, dated May 23, 2014, through an official Freedom of Information Act request.

11. For example, HR 800: New Deal Rural Broadband Act (2017–2018) and S. 454: Office of Rural Broadband Act (2019).

12. The program's title is somewhat oxymoronic, given that it is meant to connect many communities for the first time.

13. The RUS falls under the USDA's Office of Rural Development.

14. The New Deal Rural Broadband Act (2017–2018), introduced by Rep. Jared Huffman (D-CA), promised to recover the successes of the REA by establishing an office of rural broadband at the USDA and providing addition funding. Unfortunately, no action has been taken on this bill.

Chapter 4

1. "Holler" is a local colloquialization of the word "hollow," meaning "valley."

2. See the Community Network Map assembled by Community Networks (2020), a project of the ILSR.

3. The NTCA (2018) represents 850 local telecommunications companies, including telephone cooperatives. Of these members, all but 0.5 percent offer broadband, with

91.2 percent offering fiber to the home (at least for some customers); 65.8 percent rely at least in part on a copper loop, and 37.3 percent offer fiber to the node.

4. Dozens of electric cooperatives also provide high-speed broadband, although their names are more place based than people based: Co-Mo Electric, Maquoketa Valley Electric, or, in my home state of Virginia, the Bath, Alleghany, and Rockbridge (BARC) Electric Cooperative.

5. For such contemplations, see Ali (2017a), Braman (2007), Wilken (2011), and Dirlik (1999).

6. While an entire history of municipal broadband is beyond the scope of this chapter, it is a history worth noting. See Dunne (2007), Kruger and Gilroy (2016), and Muninetworks.org.

7. State of Tennessee, State of North Carolina v. Federal Communications Commission, No. 15-3291 (6th Cir. 2016).

8. It can be argued that municipal broadband actually encourages a positive reading of the First Amendment, which suggests "freedom to" provide venues for speech, rather than the reigning negative reading, which suggests "freedom from" government intervention.

9. Landgraf (2020) qualifies his findings by noting that duopoly markets do lead to an increase in DSL speeds. The same is not found for cable ISPs.

10. The GTFB (2012) reported that Woodstock "encountered $4–6 [million] in higher than project fiber and wage costs" (41).

11. RS Fiber ran up a deficit in 2018–2019, putting taxpayers on the hook for about $1 million. This means that community members will see taxes rise about $9 per month. But, as Mark Erickson, architect of RS Fiber rebutted, "For me, one of the nice side benefits of RS Fiber is the money I have saved. And while my taxes might go up $9 a month, I am saving $40 on my monthly cable and Internet bill. In fact, I have saved about $1,200 in the 30 months I've been an RS Fiber customer" (qtd. in Treacy 2018a).

12. Reisch also serves on the state's broadband task force.

13. According to the NRECA, "The smart grid consists of digital technologies, including sensors, controls, advanced meters, computers, automation, and communications, working together to optimize utility operations. A broadband backbone communications system connects this critically important grid infrastructure and can accommodate the increasingly data intensive information flows essential to modern-day grid operations" (Tucker, Goodenbery, and Loving 2018, 9).

14. For more on Dillon's Rule, see National League of Cities (2016).

Chapter 5

1. Such as the Future of Farming House Hearing (2017), High-Tech Agriculture House Hearing (2017), Examining the Farm Economy Senate Hearing (2017), and Big Data and Agriculture House Hearing (2015).

2. Some resist this term. Siva Vaidhyanathan (2019), for instance, believes it a misnomer, since the IoT is about people, not things. Laura DeNardis (2020) prefers the term "cyber-physical systems" since it better captures the range of devise and applications within this technological ecosystem.

3. A combine, or combine harvester, integrates three harvesting operations: reaping, threshing, and winnowing.

4. Deere (2015b) says that they expect transmission rates to fall to 50 percent in the coming years, adding that "this problem cannot be resolved by relying on satellite services or even more spectrum" (4).

5. Deere is not the only organization seeking control over the broadband map. USTelecom—the lobbying association for American telecommunications companies— has also sought to direct the map by recommending a new polygon-based mapping methodology that it tested in Virginia and Missouri (Stegeman 2019).

6. *United States of America v. Deere & Company, Precision Planting LLC and Monsanto Company* (N.D. Ill. 2016).

7. Bayer, which bought Monsanto in 2018, holds 29 percent of the global seed market and 25 percent of the global agrochemical market. Dow DuPont (now called Corteva) has 18 percent of the global seed market and 16 percent of the global agrochemical market. Syngenta, a Chinese-held company, controls 9 percent of the seed market and 29 percent of the agrochemical market (M. Hendrickson et al. 2019). As Mary Hendrickson et al. (2019) summarize, "Six global firms (Monsanto, DuPont, Sygenta, Bayer, Dow and BASF) [are] estimated to control three-quarters of all private sector plant breeding research, nearly three-fifths of the commercial seed market, and over three-quarters of global agrochemical sales" (25–28).

Conclusion

1. H.B. 2664 – 2017–18: Extending existing telecommunications authority to all ports in Washington State in order to facilitate public-private partnerships in wholesale telecommunications services and infrastructure.

2. The 2021 Consolidated Appropriations Act allocated $7 billion for broadband access. This included, among other provisions, $3.2 billion for an Emergency Broadband Benefit Program that will subsidize connectivity for low-income households at $50/month ($75/month on Tribal lands), $1 billion for Tribal connectivity and $300

million to the NTIA to be distributed as grants to state/local/provider partnerships for rural broadband deployment (Consolidated Appropriations Act 2021). Notwithstanding the NTIA provision, the primary focus of the broadband component of the act was clearly access rather than deployment.

3. This is a conservative estimate. CTIA, the telecommunications lobbying association, suggests that small-cell deployment could cost upward of $33,000 per cell when all of the necessary reviews are taken into account (Accenture Strategy 2018).

4. The protagonists of Samuel Beckett's *Waiting for Godot* ([1954] 2011) talk in circles as they spend the entire play waiting for someone who never comes.

5. Title IX Section 904 of the 2021 Consolidated Appropriations Act included the Broadband Interagency Coordination Act of 2020, which orders the FCC, USDA, and NTIA to enter into an interagency coordination agreement with the FCC as the lead agency. As noted in chapter 3, this is not the first time a call for interagency coordination has been made.

6. In 2019, Senators Amy Klobuchar (D-MN) and Kevin Cramer (R-ND) introduced the Office of Rural Broadband Act into the Senate. If passed, the act would establish an Office of Rural Broadband at the FCC.

7. During the revision process of this book, numerous calls came out to reform the USF. These included internal calls within the USAC to automate certain review processes (Boyd 2020) and calls from AT&T to abandon the USF completely and have rural broadband rely entirely on congressional appropriations (J. Marsh 2020). I argued against AT&T's proposal in an August 2020 op-ed with Harin Contractor (Ali and Contractor 2020).

References

Accenture Strategy. 2018. *Impact of Federal Regulatory Reviews on Small Cell Deployment*. Ann Arbor, MI: Accenture Strategy. https://api.ctia.org/wp-content/uploads/2018/04/Accenture-Strategy-Impact-of-Federal-Regulatory-Reviews-On-Small-Cell-Deployment-Report_2018.pdf.

Accessible, Affordable Internet for All Act. H.R. 7302. 116th Cong. https://www.congress.gov/bill/116th-congress/house-bill/7302.

Adler, Richard. 2015. "The Internet of Growing Things." *Computerworld*, June 4. https://www.computerworld.com/article/2931815/internet/the-internet-of-growing-things.html.

ADTRAN, Inc. 2019. "Comments of ADTRAN, Inc." (WC Docket Nos. 19-126, 10-90). Federal Communications Commission. https://ecfsapi.fcc.gov/file/10920244720800/ADTRAN%20RDOF%20Comments%209-20-19.pdf.

Agriculture Improvement Act of 2018. H.R. 2. 115th Cong. Pub. L. No. 115-334. https://www.congress.gov/bill/115th-congress/house-bill/2.

Agris Co-operative. 2019. "Farm Data so Local, It's Personalized." *West Elgin Chronicle*, March 27. https://shopping.thechronicle-online.com/places/view/362/agris_co_operative.html.

Agyeman, Julian, and Rachel Spooner. 1997. "Ethnicity and the Rural Environment." In *Contested Countryside Cultures: Otherness, Marginalisation and Rurality*, edited by P. Cloke and J. Little, 197–217. London: Routledge.

Akin Gump Strauss Hauer & Feld. 2020. "Broadband Stimulus and Related Initiatives in the HEROES Act." May 14. https://www.akingump.com/en/news-insights/heroes-act-broadband-summary.html.

Alba, Davey, and Cecilia Kang. 2020. "So We're Working from Home. Can the Internet Handle It?" *New York Times*, March 16. https://www.nytimes.com/2020/03/16/technology/coronavirus-working-from-home-internet.html.

Ali, Christopher. 2016. "The Merits of Merit Goods: Local Journalism and Public Policy in a Time of Austerity." *Journal of Information Policy* 6: 105–128. https://doi.org/10.5325/jinfopoli.6.2016.0105.

Ali, Christopher. 2017a. *Media Localism: The Policies of Place*. Champaign: University of Illinois Press.

Ali, Christopher. 2017b. "Regulatory (de) Convergence: Localism, Federalism, and Nationalism in American Telecommunications Policy." In *Media Convergence and Deconvergence*, edited by Sergio Sparviero, Corinna Peil, and Gabriele Balbi, 285–304. Cham, Switzerland: Palgrave Macmillan.

Ali, Christopher. 2018. *Thoughts on a Critical Theory of Rural Communication* (Center for Advanced Research in Global Communication Paper 7). Pennsylvania: University of Pennsylvania, Annenberg School for Communication. https://www.asc.upenn.edu/sites/default/files/documents/CARGC-Paper%207%20for%20web.pdf.

Ali, Christopher. 2019a. "An Office of Rural Broadband: We've Heard This Before." Benton Institute for Broadband and Society. March 18. https://www.benton.org/blog/office-rural-broadband-we%E2%80%99ve-heard.

Ali, Christopher. 2019b. "Thoughts on Rural Broadband Subsidies for the New Decade." Benton Institute for Broadband and Society. December 18. https://www.benton.org/blog/thoughts-rural-broadband-subsidies-new-decade.

Ali, Christopher. 2019c. "We Need a National Rural Broadband Plan." *New York Times*, February 11. https://www.nytimes.com/2019/02/06/opinion/rural-broadband-fcc.html.

Ali, Christopher. 2020. "The Politics of Good Enough: Rural Broadband and Policy Failure in the United States." International Journal of Communication 14: 5982–6004. https://ijoc.org/index.php/ijoc/article/view/15203/3285.

Ali, Christopher, and Harin Contractor. 2020. "Closing the Digital Divide Requires a Coalition on Reform of the Universal Service Fund." *The Hill*, August 17. https://thehill.com/opinion/technology/512339-closing-the-digital-divide-requires-a-coalition-on-reform-of-the-universal.

Ali, Christopher, and Mark Duemmel. 2018. "The Reluctant Regulator: The Rural Utilities Service and American Broadband Policy." *Telecommunications Policy* 43, no. 4: 380–392. https://doi.org/10.1016/j.telpol.2018.08.003.

Ali, Christopher, and Christian Herzog. 2018. "From Praxis to Pragmatism: Junior Scholars and Policy Impact." *Communication Review* 22, no. 4: 249–270. https://doi.org/10.1080/10714421.2018.1492284.

Ali, Christopher, and Manuel Puppis. 2018. "When the Watchdog Neither Barks nor Bites: Communication as a Power Resource in Media Policy and Regulation." *Communication Theory*, 28, no. 3: 270–291. https://doi.org/10.1093/ct/qtx003.

Ali, Christopher, Damian Radcliffe, Thomas R. Schmidt, and Rosalind Donald. 2018. "Searching for Sheboygans: On the Future of Small Market Newspapers." *Journalism* 21, no. 4: 453–471. https://doi.org/10.1177/1464884917749667.

Alleven, Monica. 2020a. "FCC Gives WISPs Access to 5.9 GHz for COVID-19 Response." *FierceWireless*, March 27. https://www.fiercewireless.com/wireless/fcc-gives-wisps-access -to-5-9-ghz-for-covid-19-response.

Alleven, Monica. 2020b. "Verizon, Dish & Cable Top List of CBRS Auction Winners." *FierceWireless*, September 2. https://www.fiercewireless.com/operators/verizon -dish-cable-top-list-cbrs-auction-winners.

Alliance for Affordable Internet. 2020. "Rural Broadband Policy Framework: Connecting the Unconnected." February 25. https://a4ai.org/rural-broadband-policy-frame work-connecting-the-unconnected.

Aman, Sally. 2017. "Dig Once: A Solution for Rural Broadband." USTelecom. April 12. https://www.ustelecom.org/blog/dig-once-solution-rural-broadband.

American Farm Bureau Federation. 2009. "Comments of the American Farm Bureau Federation re: GN Docket No. 09-29." Federal Communications Commission. https://www.fcc.gov/ecfs/filing/5515348661.

American Farm Bureau Federation. 2016a. "Farm Bureau Survey: Farmers Want to Control Their Own Data." May 11. https://www.fb.org/newsroom/farm-bureau-survey -farmers-want-to-control-their-own-data.

American Farm Bureau Federation. 2016b. "Privacy and Security Principles for Farm Data." https://www.fb.org/issues/innovation/data-privacy/privacy-and-security-princi ples-for-farm-data.

American Hospital Association. 2016. *Task Force on Ensuring Access in Vulnerable Communities*. Chicago: American Hospital Association. https://www.aha.org/system /files/content/16/ensuring-access-taskforce-report.pdf.

American Recovery and Reinvestment Act of 2009. 111th Cong. Pub. L. 111-5. https://www.congress.gov/bill/111th-congress/house-bill/1/text.

Anderson, Monica. 2018. "For 24% of Rural Americans, High-Speed Internet Is a Major Problem." *FactTank: News in the Numbers*, September 10. https://www.pewresearch.org /fact-tank/2018/09/10/about-a-quarter-of-rural-americans-say-access-to-high-speed -internet-is-a-major-problem.

Anderson, Monica, and Madhumitha Kumar. 2019. "Digital Divide Persists Even as Lower-Income Americans Make Gains in Tech Adoption." *FactTank: News in the Numbers*, May 7. https://www.pewresearch.org/fact-tank/2019/05/07/digital-divide -persists-even-as-lower-income-americans-make-gains-in-tech-adoption.

Anderson, Monica, and Andrew Perrin. 2018. "Nearly One-in-Five Teens Can't Always Finish Their Homework Because of the Digital Divide." *FactTank: News in the Numbers*, October 26. https://www.pewresearch.org/fact-tank/2018/10/26 /nearly-one-in-five-teens-cant-always-finish-their-homework-because-of-the-digital -divide.

Anderson, Monica, Andrew Perrin, Jinging Jiang, and Madhumitha Kumar. 2019. "10% of Americans Don't Use the Internet. Who Are They?" *FactTank: News in the Numbers*, April 22. https://www.pewresearch.org/fact-tank/2019/04/22/some-americans-I-use -the-internet-who-are-they.

Andrejevic, Mark. 2007. "Surveillance in the Digital Enclosure." *Communication Review* 10, no. 4: 295–317. https://doi.org/10.1080/10714420701715365.

Arbel, Tali, and Michael Casey. 2020. "Those without Broadband Struggle in a Stay-at-Home Nation." *PBS News Hour*, March 31. https://www.pbs.org/newshour/nation /those-without-broadband-struggle-in-a-stay-at-home-nation.

Ashmore, Fiona, John Farrington, and Sarah Skerratt. 2017. "Community-Led Broadband in Rural Digital Infrastructure Development: Implications for Resilience." *Journal of Rural Studies* 54: 408–425. https://doi.org/10.1016/j.jrurstud.2016.09.004.

Association of Equipment Manufacturers. 2017. "How Smart Farms Are Making the Case for Rural Broadband." *Growing Georgia*, October 23. https://georgia.growingamerica .com/news/2017/10/how-smart-farms-are-making-case-rural-broadband.

Atkinson, Paul. 2007. *Ethnography: Principles in Practice*. 3rd ed. New York: Routledge.

Auxier, Brooke, and Monica Anderson. 2020. "As Schools Close Due to the Coronavirus, Some U.S. Students Face a Digital 'Homework Gap.'" *FactTank: News in the Numbers*, March 16. https://www.pewresearch.org/fact-tank/2020/03/16/as -schools-close-due-to-the-coronavirus-some-u-s-students-face-a-digital-homework -gap/.

Baker, C. Edwin. 2002. *Media, Markets, and Democracy*. Cambridge: Cambridge University Press.

Baldwin, Robert, Martin Cave, and Martin Lodge. 2012. *Understanding Regulation: Theory, Strategy, and Practice*. 2nd ed. Oxford: Oxford University Press.

Baller Stokes & Lide. 2019. "State Restrictions on Community Broadband Services or Other Public Communications Initiatives." July 1. https://www.baller.com/2018/01 /state-restrictions-on-community-broadband-services-or-other-public-communications -initiatives/.

Band, Jonathan. 2018. "Expanded DMCA Exemptions Enhance Competition and Innovation." Disruptive Competition Project. October 30. http://www.project-disco .org/intellectual-property/103018-expanded-dmca-competition-innovation.

Baran, Paul A., and Paul M. Sweezy. 1966. *Monopoly Capital: An Essay on the American Economic and Social Order*. New York: NYU Press.

Bargues-Pedreny, Pol, David Chandler, and Elena Simon. 2019. "Mapping and Politics in the Digital Age: An Introduction." In *Mapping and Politics in the Digital Age*, edited by Pol Bargues-Pedreny, David Chandler, and Elena Simon, 1–19. New York: Routledge.

Barney, Darin. 2011. "To Hear the Whistle Blow: Technology and Politics on the Battle River Branch Line." *TOPIA: Canadian Journal of Cultural Studies* 25. https://utpjournals.press/doi/abs/10.3138/topia.25.5.

Bartholomew, Darin. 2015. "Long Comment regarding a Proposed Exemption under 17 U.S.C. 1201: Deere & Company." https://cdn.loc.gov/copyright/1201/2015/comments-032715/class%2021/John_Deere_Class21_1201_2014.pdf.

Bass, Jim. 2010. *NTCA: A History*. Arlington, VA: National Telecommunications Cooperative Association.

Bator, Francis. M. 1958. "The Anatomy of Market Failure." *Quarterly Journal of Economics* 72, no. 3: 351–379.

Bauerly, Brittney, Russell McCord, Rachel Hulkower, and Dawn Pepin. 2019. "Broadband Access as a Public Health Issue: The Role of Law in Expanding Broadband Access and Connecting Underserved Communities for Better Health Outcomes." *Journal of Law, Medicine & Ethics* 47, no. 2: 39–42. https://doi.org/10.1177/1073110519857314.

Baumgartner, Jeff. 2018. "Multi-Gig Brings Star Trek Holodeck to Cable." *Light-Reading*, August 9. https://www.lightreading.com/video/multi-gig-brings-star-trek-holodeck-to-cable/a/d-id/745272.

BBC. 2010. "Finland Makes Broadband a 'Legal Right,'" July 1. https://www.bbc.com/news/10461048.

Beckett, Samuel. [1954] 2011. *Waiting for Godot: A Tragicomedy in Two Acts*. New York: Grove Press.

Bell, Michael M., Sarah E. Lloyd, and Christine Vatovec. 2010. "Activating the Countryside: Rural Power, the Power of the Rural and the Making of Rural Politics." *Sociologia Ruralis* 50, no. 3, 205–224. https://doi.org/10.1111/j.1467-9523.2010.00512.x.

Bennett, Kevin J., Matthew Yuen, and Francisco Blanco-Silva. 2018. "Geographic Differences in Recovery after the Great Recession." *Journal of Rural Studies* 59: 111–117. https://doi.org/10.1016/j.jrurstud.2018.02.008.

Bentley, Kipp. 2018. "School Buses Become Wi-Fi Hot Spots." *GovTech*, June 13. https://www.govtech.com/education/k-12/School-Buses-Become-WiFi-Hot-Spots.html.

Berdik, Chris. 2018. "Rural Kids Face an Internet 'Homework Gap.' The FCC Could Help." *Wired*, November 12. https://www.wired.com/story/rural-kids-internet-homework-gap-fcc-could-help.

Berman, David Elliot. "We Need Broadband Internet for All." *Jacobin*, December 23. https://www.jacobinmag.com/2019/12/broadband-internet-bernie-sanders.

Berry, Wendell. 2006. *The Way of Ignorance and Other Essays*. Berkeley, CA: Counterpoint.

Bevans, Stephen. B. 2002. *Models of Contextual Theology*. Rev. ed. Maryknoll, NY: Orbis Books.

Black, Julia. 2008. "Constructing and Contesting Legitimacy and Accountability in Polycentric Regulatory Regimes." *Regulation & Governance* 2, no. 2: 137–164. https://doi.org/10.1111/j.1748-5991.2008.00034.x.

Bloch, Sam. 2018. "If Farmers Sold Their Data Instead of Giving It Away, Would Anybody Buy?" *The Counter*, July 19. https://thecounter.org/farmobile-farm-data.

Bode, Karl. 2018a. "FCC Falsely Claims Community Broadband an 'Ominous Threat to the First Amendment.'" *Vice*, October 29. https://www.vice.com/en_us/article/bj49j8/fcc-falsely-claims-community-broadband-an-ominous-threat-to-the-first-amendment.

Bode, Karl. 2018b. "How Bad Maps Are Ruining American Broadband." *The Verge*, September 24. https://www.theverge.com/2018/9/24/17882842/us-internet-broadband-map-isp-fcc-wireless-competition.

Boyd, Aaron. 2020. "The Universal Service Administrative Company Is Looking at Robotic Process Automation to Streamline Checks and Balances before It Subsidizes Telephone and Internet Access in High-Cost Areas." *Nextgov*, August 12. https://www.nextgov.com/emerging-tech/2020/08/universal-service-fund-wants-automate-verification-high-cost-telecom-broadband-areas/167653/.

Braman, Sandra. 2007. "The Ideal vs. the Real in Media Localism: Regulatory Implications." *Communication Law and Policy* 12, no. 3: 231–278.

Braman, Sandra. 2009. *Change of State: Information, Policy, and Power*. Cambridge, MA: MIT Press.

Bray, Hiawatha. 2020. "Yes, the Internet Can Handle the Coronavirus Traffic Jam, with Hiccups." *Boston Globe*, March 18. https://www.bostonglobe.com/2020/03/18/business/can-information-superhighway-handle-coronavirus-traffic-jam/.

Britt, Phil. 2020. "Report: Business Hours Broadband Consumption Soars during Pandemic."

Telecompetitor, March 19. https://www.telecompetitor.com/report-business-hours-broadband-consumption-soars-during-pandemic/.

Broadband DATA Act of 2020. 116th Cong. Pub. L. 116-130. https://www.congress.gov/bill/116th-congress/senate-bill/1822?q=%7B%22search%22%3A%5B%22Broadband+DATA+Act%22%5D%7D&s=1&r=2.

BroadbandNow. 2019a. "Digital Divide: Broadband Pricing by State, Zip Code, and Income 2019." March 27. https://broadbandnow.com/research/digital-divide -broadband-pricing-state-zip-income-2019.

BroadbandNow. 2019b. "Internet Providers in Luverne, MN." https://broadbandnow .com/Minnesota/Luverne.

Broadband Tech Report. 2019. "Kagan: U.S. Wireline Broadband Growth Slowing." *Broadband Technology Report*, August 14. https://www.broadbandtechreport.com /docsis/article/14038350/kagan-us-wireline-broadband-growth-slowing.

Brodkin, Jon. 2015. "Want Fiber Internet? That'll Be $383,500, ISP Tells Farm Owner." *Ars Technica*, August 11. https://arstechnica.com/information-technology /2015/08/want-fiber-internet-thatll-be-383500-isp-tells-farm-owner.

Brodkin, Jon. 2017a. "8,500 Verizon Customers Disconnected Because of "Substantial" Data Use." *Ars Technica*, September 15. https://arstechnica.com/information -technology/2017/09/hristo-kicks-8500-rural-customers-off-network-for-using -roaming-data.

Brodkin, Jon. 2017b. FCC's Latest Gift to Telcos Could Leave Americans with Worse Internet Access. *Ars Technica*. https://arstechnica.com/tech-policy/2017/11/fccs -latest-gift-to-telcos-could-leave-americans-with-worse-internet-access/.

Brodkin, Jon. 2017c. "If FCC Gets Its Way, We'll Lose a Lot More Than Net Neutrality." *Ars Technica*, July 12. https://arstechnica.com/tech-policy/2017/07/how-title-ii -goes-beyond-net-neutrality-to-protect-internet-users-from-isps.

Brodkin, Jon. 2019. "Millimeter-Wave 5G Isn't for Widespread Coverage, Verizon Admits." *Ars Technica*, April 23. https://arstechnica.com/information-technology /2019/04/millimeter-wave-5g-isnt-for-widespread-coverage-verizon-admits.

Brodkin, Jon. 2020a. "Ajit Pai Caves to SpaceX but Is Still Skeptical of Musk's Latency Claims." *Ars Technica*, June 11. https://arstechnica.com/tech-policy/2020/06/ajit-pai -caves-to-spacex-but-is-still-skeptical-of-musks-latency-claims/.

Brodkin, Jon. 2020b. "AT&T's DSL Phaseout Is Leaving Poor, Rural Users Behind." Wired, October 7. https://www.wired.com/story/atandts-dsl-phaseout-is-leaving-poor -rural-users-behind/.

Brodkin, Jon. 2020c. "CenturyLink, Frontier Took FCC Cash, Failed to Deploy All Required Broadband." *Ars Technica*, January 23. https://arstechnica.com/tech-policy /2020/01/centurylink-frontier-took-fcc-cash-failed-to-deploy-all-required-broadband.

Brodkin, Jon. 2020d. "FCC Helps Charter Avoid Broadband Competition." *Ars Technica*, June 27. https://arstechnica.com/tech-policy/2020/06/fcc-helps-charter-avoid -broadband-competition/.

Brodkin, Jon. 2020e. "Frontier, Amid Bankruptcy, Is Suspected of Lying about Broadband Expansion." *Ars Technica*, May 2. https://arstechnica.com/tech-policy/2020/05/frontier-amid-bankruptcy-is-suspected-of-lying-about-broadband-expansion/.

Brown, Deward Clayton. 1980. *Electricity for Rural America: The Fight for the REA*. Westport, CT: Greenwood Press.

Buckley, Linda. 1999. *Builder of the Past, Architect of the Future: The History of the REA/RUS Telephone Program*. Arlington, VA: Foundation for Rural Service.

Bunge, Jacob. 2014. "DuPont Sees $500 Million in Annual Revenue from Farm-Data Services." *Wall Street Journal*, February 27. https://www.wsj.com/articles/dupont-sees-500-million-in-annual-revenue-from-farm-data-services-1393515676.

Busby, John, and Julia Tanberk. 2020. "FCC Underestimates Americans Unserved by Broadband Internet by 50%." BroadbandNow. February 3. https://broadbandnow.com/research/fcc-underestimates-unserved-by-50-percent.

Busch, Fritz. 2018. "Taxpayers Asked to Pony Up for RS Fiber Shortfall." *The Journal*, December 2. https://www.nujournal.com/news/local-news/2018/12/02/taxpayers-asked-to-pony-up-for-rs-fiber-shortfall/.

Cairncross, Frances. 1997. *The Death of Distance: How the Communications Revolution Is Changing Our Lives*. Cambridge, MA: Harvard University Press.

Canadian Radio-Television and Telecommunications Commissioner. 2016. *Modern Telecommunications Services: The Path Forward for Canada's Digital Economy* (Regulatory Policy No. 2016-496). December 21. Gatineau, Quebec: Canadian Radio-Television and Telecommunications Commissioner. https://crtc.gc.ca/eng/archive/2016/2016-496.pdf.

Canadian Radio-Television and Telecommunications Commission. 2020. "What You Should Know about Internet Speeds." https://crtc.gc.ca/eng/internet/performance.htm.

Carlson, Scott, and Christopher Mitchell. 2016. *RS Fiber: Fertile Fields for New Rural Internet Cooperative*. Institute for Local Self Reliance. https://ilsr.org/report-mn-rural-fiber/.

Casey, Edward. 2013. *The Fate of Place: A Philosophical History*. Berkeley: University of California Press.

Castaneda, Vera. 2020. "Spectrum and Other Providers Offer Free Internet for Students at Home." *Los Angeles Times*, March 20. https://www.latimes.com/socal/glendale-news-press/news/story/2020-03-20/spectrum-and-other-internet-providers-offer-free-internet-for-students-at-home.

Castells, Manuel. 2010. *The Rise of the Network Society*. Malden, MA: Blackwell.

Castells, Manuel. 2013. *Communication Power*. Oxford: Oxford University Press.

CCG Consulting. 2014. "Reply Comments of CCG Consulting" (WCB Dockets 14-115 and 14-116). National Exchange Carrier Association. https://prodnet.www.neca.org/publicationsdocs/wwpdf/93014ccg.pdf.

Chamberlain, Kendra. 2019. "Municipal Broadband Is Roadblocked or Outlawed in 26 States." BroadbandNow. April 17. https://broadbandnow.com/report/municipal -broadband-roadblocks.

Chambers, Jonathan. 2018. "Proving a Negative: The RUS Has New Funds to Spend on Connecting Unserved Areas. But How Can Applicants Prove an Area Is Unserved? *Broadband Communities*, August–September. http://www.bbpmag.com/Features/0818 -Proving-a-Negative.php.

Chambers, Jonathan. 2019. "The Rural Digital Opportunity Fund: Conexon's Comments Filed with the FCC." Conexon. September 23. https://www.conexon.us/conexon -blog/the-rural-digital-opportunity-fund-conexons-comments-filed-with-the-fcc.

Chao, Becky, and Claire Park. 2020. *The Cost of Connectivity 2020*. Washington, DC: Open Technology Institute. https://www.newamerica.org/oti/reports/cost-connectivity -2020/.

Charles, Dan. 2014. "Should Farmers Give John Deere and Monsanto Their Data?" *NPR*, January 22. https://www.npr.org/sections/thesalt/2014/01/21/264577744/should -farmers-give-john-deere-and-monsanto-their-data.

Chiaraviglio, Luca, Nicola Blefari-Melazzi, William Liu, Jairo A. Gutierrez, Jaap Van De Beek, Robert Birke, Lydia Chen, Filip Idzikowski, Daniel Kilper, J. Paolo Monti, and Jinsong Wu. 2016. "5G in Rural and Low-Income Areas: Are We Ready?" In *Proceedings of the ITU Kaleidoscope: ICTs for a Sustainable World (ITU WT), Bangkok, November 14–16, 2016*, 1–8. New York: IEEE.

Chiou, Lesley, and Catherine Tucker. 2020. "Social Distancing, Internet Access and Inequality" (Working Paper 26982). Cambridge, MA: National Bureau of Economic Research. https://www.nber.org/system/files/working_papers/w26982/w26982.pdf.

Chiu, Allyson. 2019. "'Queen of Condescending Applause': Nancy Pelosi Clapped at Trump, and the Internet Lost It." *Washington Post*, February 6. https://www .washingtonpost.com/nation/2019/02/06/queen-condescending-applause-nancy -pelosi-clapped-trump-internet-lost-it.

Cirani, Simone, Gianluigi Ferrari, Marco Picone, and Luca Veltri. 2018. *Internet of Things: Architectures, Protocols and Standards*. Hoboken, NJ: John Wiley and Sons.

Cobb, John T. 2018. "Broad-Banned: The FCC's Preemption of State Limits on Municipal Broadband and the Clear Statement Rule." *Emory Law Journal* 68, no. 2: 407–439.

Cobb, Stephen. 2011. "Satellite Internet Connection for Rural Broadband." Community Networks. May 2. https://muninetworks.org/reports/satellite-internet-connection-rural -broadband.

Coleman, Bill. 2018. *Impact of CAF II-Funded Networks*. Grand Rapids, MN: Blandin Foundation. https://blandinfoundation.org/learn/research-rural/broadband-resources /broadband-initiative/impact-of-caf-ii-funded-networks.

Columbus, Louis. 2018. "2018 Roundup of Internet of Things Forecasts and Market Estimates." *Forbes*, December 13. https://www.forbes.com/sites/louiscolumbus/2018/12/13/2018-roundup-of-internet-of-things-forecasts-and-market-estimates.

Communications Act of 1934 as Amended by the 1996 Telecommunications Act. Pub. L. No. 104-104, 47 U.S. Code. https://transition.fcc.gov/Reports/1934new.pdf.

Community Networks. 2020. Community Network Map. https://muninetworks.org/communitymap.

Compaine, Benjamin, ed. 2001. *The Digital Divide: Facing a Crisis or Creating a Myth?* Cambridge, MA: MIT Press.

Congressional Research Service. 2012. *The Davis-Bacon Act and Changes in Prevailing Wage Rates, 2000 to 2008* (No. R40663). Washington, DC: Congressional Research Service. https://www.everycrsreport.com/files/20120330_R40663_64b5c1083a7eb15667bcfb25816d965052602f67.pdf.

Congressional Research Service. 2019a. *Broadband Internet Access and the Digital Divide: Federal Assistance Programs* (No. RL30719). Washington, DC: Congressional Research Service. https://crsreports.congress.gov/product/pdf/RL/RL30719.

Congressional Research Service. 2019b. *U.S. Farm Income Outlook: August 2019 Forecast* (No. R45924). Washington, DC: Congressional Research Service. https://www.everycrsreport.com/files/20190919_R45924_428b4fec19f734a8d6988fc5b6fd2019f53c1f6d.pdf.

Congressional Research Service. 2020. *USDA's ReConnect Broadband Pilot Program.* Washington, DC: Congressional Research Service. https://crsreports.congress.gov/product/pdf/IF/IF11262.

Consolidated Appropriations Act of 2018. H.R. 1625. 115th Cong. Pub. L. No. 115-141. https://www.congress.gov/bill/115th-congress/house-bill/1625/text.

Consolidated Appropriations Act of 2021. H.R. 133. 116th Cong. Pub. L. No. 116-260. https://www.congress.gov/bill/116th-congress/house-bill/133.

Consumer Federation of America and Consumers Union. 2009. "Comments of the Consumer Federation of America and Consumers Union" (GN Docket No. 09-29). Federal Communications Commission. https://ecfsapi.fcc.gov/file/6520203357.pdf.

Cooke, Morris. L. 1948. "The Early Days of the Rural Electrification Idea: 1914–1936." *American Political Science Review* 42, no. 3: 431–447. https://doi.org/10.2307/1949909.

Cooper, Tyler. 2018a. "FCC Concludes Satellite Internet Is Good Enough for Rural Broadband." BroadbandNow. February 16. https://broadbandnow.com/report/satellite-internet-good-enough-rural-broadband.

Cooper, Tyler. 2018b. "WISPs Are the Real Heroes in Bridging the Digital Divide." Broad-bandNow. January 26. https://broadbandnow.com/report/wisps-real-heroes-bridging -digital-divide.

Cooper, Tyler. 2019. "Bezos and Musk's Satellite Internet Could Save Americans $30B a Year." *Podium*, August 24. https://thenextweb.com/podium/2019/08/24/bezos-and -musks-satellite-internet-could-save-americans-30b-a-year.

Copeland, Basil and Alan Severn. 1985. "Price Theory and Telecommunications Regulation: A Dissenting View." *Yale Journal on Regulation* 3: 53–85.

Copps, Michael J. 2009. *Bringing Broadband to Rural America: Report on Rural Broadband Strategy*. Washington, DC: Federal Communications Commission. https://apps .fcc.gov/edocs_public/attachmatch/DOC-291012A1.pdf.

Copyright Office. 2015. "Exemption to Prohibition on Circumvention of Copyright Protection Systems for Access Control Technologies" (37 CFR Part 201 Docket No. 2014-07). *Federal Register* 80, no. 208: 65944. https://www.copyright.gov/fedreg/2015 /80fr65944.pdf.

Copyright Office. 2018. "Exemption to Prohibition on Circumvention of Copyright Protection Systems for Access Control Technologies" (37 CFR Part 201 Docket No. 2017-10). *Federal Register* 83, no. 208: 54010. https://www.govinfo.gov/content/pkg /FR-2018-10-26/pdf/2018-23241.pdf.

Cornes, Richard, and Todd Sandler. 1996. *The Theory of Externalities, Public Goods and Club Goods*. Cambridge: Cambridge University Press.

Coronavirus Aid, Relief, and Economic Security Act of 2020. H.R. 748 116th Cong. Pub. L. No. 116-136. https://www.congress.gov/bill/116th-congress/house-bill/748.

Crampton, Jeremy. W. 2004. *The Political Mapping of Cyberspace*. Chicago: University of Chicago Press.

Crawford, Susan. 2019. *Fiber: The Coming Tech Revolution—and Why America Might Miss It*. New Haven, CT: Yale University Press.

Cromartie, John, and Shawn Bucholtz. 2008. "Defining the 'Rural' in Rural America." *Amber Waves*, June 1. https://www.ers.usda.gov/amber-waves/2008/june/defining-the -rural-in-rural-america/.

CTC Technology & Energy. 2018. *A Model for Understanding the Cost to Connect Anchor Institutions with Fiber Optics*. Washington, DC: Schools, Health & Libraries Broadband (SHLB) Coalition. http://www.shlb.org/uploads/Policy/Infrastructure/SHLB_Connect ingAnchors_CostEstimate.pdf.

Curl, John. 2010. *For All the People: Uncovering the Hidden History of Cooperation, Cooperative Movements, and Communalism in America*. Oakland, CA: PM Press.

Dampier, Phillip. 2018. "Strong Evidence CenturyLink Giving Up on Most Residential Broadband Upgrades." *Stop the Cap!*, March 27. https://stopthecap.com/2018/03/26 /strong-evidence-centurylink-giving-up-on-most-residential-broadband-upgrades.

Dano, Mike. 2020. "Verizon, SpaceX, CenturyLink, Charter among FCC's RDOF Bidders." *LightReading*, September 1. https://www.lightreading.com/verizon-spacex -centurylink-charter-among-fccs-rdof-bidders/d/d-id/763599.

Davidson, Charles, and Michael Santorelli. 2014. *Understanding the Debate over Government-Owned Broadband Networks.* New York: Advanced Communications Law and Policy Institute. https://www.heartland.org/_template-assets/documents/publications /aclp-government-owned-broadband-networks-final-june-2014.pdf.

Dawson, Doug. 2018a. "A-CAM: A Subsidy that Works." *POTs and PANs*, January 31. https://potsandpansbyccg.com/2018/01/31/a-cam-a-subsidy-that-works.

Dawson, Doug. 2018b. "Big Telcos and Rural Customers." *POTs and PANs*, January 22. https://potsandpansbyccg.com/2018/01/22/big-telcos-and-rural-customers.

Dawson, Doug. 2018c. "Getting Militant for Broadband." *POTs and PANs*, January 9. https://potsandpansbyccg.com/2018/07/09/getting-militant-for-broadband.

Dawson, Doug. 2020. "A New Rural Broadband Product." *POTS and PANS*, August 10. https://potsandpansbyccg.com/tag/jetpack/.

Deere & Company. 2014a. "Comments of Deere & Company." (WC Docket No. 10-90). Federal Communications Commission. https://www.fcc.gov/ecfs/filing/6018254513.

Deere & Company. 2014b. "Notice of Exparte Discussion: Infrastructure Gaps in the U.S." (WC Docket No. 19-90). Federal Communications Commission. https://www .fcc.gov/ecfs/filing/6017600514.

Deere & Company. 2015a. "Comments of Deere & Company" (GN Docket No. 14-126). National Exchange Carrier Association. https://prodnet.www.neca.org/publicationsdocs /wwpdf/91615deere.pdf.

Deere & Company. 2015b. "Reply Comments of Deere & Company" (GN Docket No. 14-126). Federal Communications Commission. https://ecfsapi.fcc.gov/file/6000104 2484.pdf.

De Sa, Paul. 2017. *Improving the Nation's Digital Infrastructure.* Washington, DC: Federal Communications Commission. https://www.fcc.gov/document/improving-nations -digital-infrastructure.

Deller, Steven, and Brian Whitacre. 2019. "Broadband's Relationship to Rural Housing Values." *Papers in Regional Science* 98, no. 5: 2135–2156. https://doi.org/10.1111/pirs .12450.

DeNardis, Laura. 2020. *The Internet in Everything: Freedom and Security in a World with No Off Switch.* New Haven, CT: Yale University Press.

de Viti de Marco, Antonio. 1936. *First Principles of Public Finance*. Translated by Edith Pavlo Marget. London: Jonathan Cape.

Digital Millennium Copyright Act of 1998. H.R. 2281. 105th Cong. Pub. L. No. 105-304. https://www.govinfo.gov/content/pkg/PLAW-105publ304/pdf/PLAW-105publ304.pdf.

Dinterman, Robert, and Mitch Renkow. 2017. "Evaluation of USDA's Broadband Loan Program: Impacts on Broadband Provision." *Telecommunications Policy* 41, no. 2: 140–153. https://doi.org/10.1016/j.telpol.2016.12.004.

Dirlik, Arif. 1999. "Place-Based Imagination: Globalism and the Politics of Place." In *Place and Politics in an Age of Globalization*, edited by Roxann Prazniak and Arif Dirlik, 15–52. Lanham, MD: Rowman & Littlefield.

DownEast Broadband Utility. 2018. *"E-Connectivity Pilot Program: Comments from Downeast Broadband"* (Docket No. RUS-18-TELECOM-004). https://beta.regulations.gov/comment/RUS-18-TELECOM-0004-0127.

Draper, Nora A. 2019. *The Identity Trade: Selling Privacy and Reputation Online*. New York: NYU Press.

Duarte, Marissa. 2017. *Network Sovereignty: Building the Internet across Indian Country*. Seattle: University of Washington Press.

Ducket, Chris. 2020. "Telstra and Ericsson Get 200km Range on LTE Signal with Software Upgrade." *ZDNet*, February 27. https://www.zdnet.com/article/telstra-and-ericsson-get-200km-range-on-lte-signal-with-software-upgrade/.

Dunbar-Hester, Christina. 2013. "What's Local? Localism as a Discursive Boundary Object in Low-Power Radio Policymaking." *Communication, Culture & Critique* 6, no. 4: 502–524. https://doi.org/10.1111/cccr.12027.

Duncan, Kevin, and Russell Ormiston. 2018. "What Does the Research Tell Us about Prevailing Wage Laws?" *Labor Studies Journal* 44, no. 2: 139–206. https://doi.org/10.1177/0160449X18766398.

Dunne, Matthew. 2007. "Let My People Go (Online): The Power of the FCC to Preempt State Laws that Prohibit Municipal Broadband." *Columbia Law Review* 107, no. 5: 1126–1163.

Early, Wesley. 2020. "FCC Opens Priority Access for Tribes to Get Broadband Licenses." *Alaska Public Media*, July 14. https://www.alaskapublic.org/2020/07/14/fcc-opens-priority-access-for-tribes-to-get-broadband-licenses/.

Eggerton, John. 2020. "Pai: FCC Needs Mapping Funding from Congress ASAP." *MultiChannel News*, March 24. https://www.multichannel.com/news/pai-fcc-needs-mapping-funding-from-congress-asap.

Engebretson, Joan. 2015. "AT&T and Verizon CAF Plans for Rural Broadband Heading in Two Different Directions." *Telecompetitor*. https://www.telecompetitor

.com/att-and-verizon-caf-plans-for-rural-broadband-heading-in-two-different-directions/.

Engebretson, Joan. 2016. "Exec: AT&T Fixed Wireless Planned for CAF-Funded Rural Areas." *Telecompetitor*, August 10. https://www.telecompetitor.com/exec-att-fixed-wireless-planned-for-caf-funded-rural-areas.

Engebretson, Joan. 2018a. "Economists Put the Tab at $61 Billion to Bring Fiber Broadband to Rural U.S." *Telecompetitor*, July 11. https://www.telecompetitor.com/economists-put-the-tab-at-61-billion-to-bring-fiber-broadband-to-rural-u-s.

Engebretson, Joan. 2018b. "How the Rural Electric Cooperative Consortium Won $186 Million in CAF II Funding for Gigabit Broadband." *Telecompetitor*, August 30. https://www.telecompetitor.com/how-the-rural-electric-cooperative-consortium-won-caf-ii-funding-for-gigabit-broadband/.

Engebretson, Joan. 2018c. "T-Mobile: One 600 MHz Cell Tower Will Cover Hundreds of Miles with 5G, but at What Speed?" *Telecompetitor*, November 20. https://www.telecompetitor.com/t-mobile-one-600-mhz-cell-tower-will-cover-hundreds-of-miles-with-5g-but-at-what-speed/.

Engebretson, Joan. 2019a. "CEO: Rural Digital Opportunity Fund May Be 'Less Favorable to Frontier' Than CAF Program Was." *Telecompetitor*, August 7. https://www.telecompetitor.com/ceo-rural-digital-opportunity-fund-may-be-less-favorable-to-frontier-than-caf-program-was.

Engebretson, Joan. 2019b. "FCC Authorizes $4.9 Billion in A-CAM Rural Broadband Funding." *Telecompetitor*, August 23. https://www.telecompetitor.com/fcc-authorizes-4-9-billion-in-a-cam-rural-broadband-funding.

Engebretson, Joan. 2019c. "T-Mobile Sprint Merger Conditions Would Include 5G Build-Out, Fixed Wireless Home Internet Commitments." *Telecompetitor*, May 20. https://www.telecompetitor.com/t-mobile-sprint-merger-conditions-would-include-5g-build-out-fixed-wireless-home-internet-commitments.

Engebretson, Joan. 2020a. "FCC Decision on 6 GHz Unlicensed Spectrum May Supercharge the Wi-Fi Industry, Fixed Wireless. *Telecompetitor*, April 23. https://www.telecompetitor.com/fcc-decision-on-6-ghz-unlicensed-spectrum-may-supercharge-the-wi-fi-industry-fixed-wireless/.

Engebretson, Joan. 2020b. RDOF Eligible Areas: Is Frontier Seeking a Loophole? *Telecompetitor*, April 29. https://www.telecompetitor.com/rdof-eligible-areas-is-frontier-seeking-a-loophole/.

Estes, Adam. 2015. "John Deere Thinks People Will Pirate Music with In-Car Computers." *Gizmodo*, April 3. https://gizmodo.com/john-deere-thinks-people-will-pirate-music-with-in-car-1695574031.

European Commission. 2019. "Commission Decides to Harmonise Radio Spectrum for the Future 5G." January 24. https://ec.europa.eu/digital-single-market/en/news /commission-decides-harmonise-radio-spectrum-future-5g.

European Commission. 2020. "Country Information—Germany." https://ec.europa .eu/digital-single-market/en/country-information-germany.

European Union. 2020. "Free Wi-Fi for Europeans." WiFi4EU Portal. https://wifi4eu .ec.europa.eu.

Fairlie, Robert W., Rebecca A. London, Rachel Rosner, and Manuel Pastor. 2006. *Crossing the Divide: Immigrant Youth and Digital Disparity in California*. Santa Cruz, CA: Center for Justice, Tolerance and Community. https://cjtc.ucsc.edu/docs/digital.pdf.

Falcon, Ernesto. 2020. "The American Federal Definition of Broadband Is Both Useless and Harmful." Electronic Frontier Foundation. July 17. https://www.eff.org/deeplinks /2020/07/american-federal-definition-broadband-both-useless-and-harmful.

Fauquier Times. 2019. "Say Goodbye to Copper Telephone Landlines," September 10. https://www.fauquier.com/news/say-goodbye-to-copper-telephone-landlines/article _28b5633e-cf4d-11e9-88f4-4b1e1f5b697c.html.

Federal Communications Commission. n.d. "Bridging the Digital Divide for All Americans." https://www.fcc.gov/about-fcc/fcc-initiatives/bridging-digital-divide-all -americans.

Federal Communications Commission. 2006. Land of Opportunity: Bringing Telecommunications Services to Rural Communities. Washington, DC: Federal Communications Commission. https://transition.fcc.gov/indians/opportunity.pdf

Federal Communications Commission. 2010. *Connecting America: The National Broadband Plan*. Washington, DC: Federal Communications Commission. https:// transition.fcc.gov/national-broadband-plan/national-broadband-plan.pdf.

Federal Communications Commission. 2011a. *Bringing Broadband to Rural America: Update to Report on a Rural Broadband Strategy*. Washington, DC: Federal Communications Commission. https://www.fcc.gov/document/bringing-broadband-rural -america.

Federal Communications Commission. 2011b. *Report and Order and Further Notice of Proposed Rulemaking* (FCC 11-161, WC Docket No. 10-90). Washington, DC: Federal Communications Commission. https://www.fcc.gov/document/fcc-releases-connect -america-fund-order-reforms-usficc-broadband.

Federal Communications Commission. 2012. *Notice of Proposed Rulemaking and Order* (FCC 12-148, GN Docket No. 12-354). Washington, DC: Federal Communications Commission. https://ecfsapi.fcc.gov/file/7022080889.pdf.

Federal Communications Commission. 2014a. Notice of Proposed Rulemaking (FCC 14-144, ET Docket No. 14-165, GN Docket No. 12-268). Washington, DC: Federal Communications Commission. https://ecfsapi.fcc.gov/file/60000870171.pdf.

Federal Communications Commission. 2014b. *Order* (DA 14-1042). Washington, DC: Federal Communications Commission. https://www.fcc.gov/document/update -rural-areas-rural-health-care-program.

Federal Communications Commission. 2014c. *Report and Order and Further Notice of Proposed Rulemaking* (FCC 14-98, WC Docket Nos. 10-90, 14-58). Washington, DC: Federal Communications Commission. https://docs.fcc.gov/public/attachments/FCC -14-98A1.pdf.

Federal Communications Commission. 2014d. *Report and Order, Declaratory Ruling, Order, Memorandum Opinion and Order, Seventh Order on Reconsideration, and Further Notice of Proposed Rulemaking* (FCC 14-54, CC Docket No. 01-92, WC Docket Nos. 10-90, 14-58, 07-135, WT Docket No. 10-208). https://www.fcc.gov/document/fcc -releases-connect-america-fund-omnibus-order-and-fnprm.

Federal Communications Commission. 2015a. "Connect America Fund Phase II Funding by Carrier, State, and County." https://www.fcc.gov/document/connect -america-fund-phase-ii-funding-carrier-state-and-county.

Federal Communications Commission. 2015b. "Wireline Competition Bureau Announces Connect America Phase II Support Amounts Offered to Price Cap Carriers to Expand Rural Broadband (WC Docket No. 10-90)." https://www.fcc.gov /document/model-based-support-offers-pn.

Federal Communications Commission. 2016. *Report and Order, Order and Order on Reconsideration, and Further Notice of Proposed Rulemaking* (FCC 16-33, CC Docket No. 01-92, WC Docket Nos. 10-90, 14-58). Washington, DC: Federal Communications Commission. https://www.fcc.gov/document/fcc-reforms-high-cost-program-rate-return-carriers.

Federal Communications Commission. 2018a. *Connect America Fund Phase II: Bids—All Bids*. Washington, DC: Federal Communications Commission. https:// auctiondata.fcc.gov/public/projects/auction903/reports/prs_all_bids.

Federal Communications Commission. 2018b. *Declaratory Ruling and Third Report and Order* (FCC 18-133, WC Docket No. 17-79, WT Docket No. 17-84). Washington, DC: Federal Communications Commission. https://docs.fcc.gov/public/attachments /FCC-18-133A1.pdf.

Federal Communications Commission. 2018c. *Declaratory Ruling, Report and Order, and Order* (FCC 17-166, WC Docket No. 17-108). Washington, DC: Federal Communications Commission. https://www.fcc.gov/document/fcc-releases-restoring-internet -freedom-order.

Federal Communications Commission. 2018d. *Eighth Measuring Broadband America Fixed Broadband Report.* Washington, DC: Federal Communications Commission. https://data.fcc.gov/download/measuring-broadband-america/2018/2018-Fixed-Measuring-Broadband-America-Report.pdf.

Federal Communications Commission. 2018e. "Fixed Broadband Deployment." https://broadbandmap.fcc.gov/#/.

Federal Communications Commission. 2018f. *Report and Order, Further Notice of Proposed Rulemaking, and Order on Reconsideration* (FCC 18-176, CC Docket No. 01-92, WC Docket Nos. 10-90, 14-58, 07-135). Washington, DC: Federal Communications Commission. https://docs.fcc.gov/public/attachments/FCC-18-176A1.pdf.

Federal Communications Commission. 2019a. *2019 Broadband Deployment Report* (FCC 19-44). Washington, DC: Federal Communications Commission. https://docs.fcc.gov/public/attachments/FCC-19-44A1.pdf.

Federal Communications Commission. 2019b. "FCC Proposes $20.4 Billion Rural Digital Opportunity Fund." August 2. https://docs.fcc.gov/public/attachments/DOC-358831A1.pdf.

Federal Communications Commission. 2019c. "Louisa County. VA." December. https://broadbandmap.fcc.gov/#/area-summary?version=dec2019&type=county&geoid=51109&tech=acfosw&speed=25_3&vlat=37.94126914635366&vlon=-77.9970285&vzoom=9.000862149652646.

Federal Communications Commission. 2019d. *Mobility Fund Phase II: Coverage Maps Investigation Staff Report* (GN Docket No. 19-367). Washington, DC: Federal Communications Commission. https://docs.fcc.gov/public/attachments/DOC-361165A1.pdf.

Federal Communications Commission. 2020a. *2020 Broadband Deployment Report.* Washington, DC: Federal Communications Commission. https://www.fcc.gov/reports-research/reports/broadband-progress-reports/2020-broadband-deployment-report.

Federal Communications Commission. 2020b. "Auction 904 Updated Eligible Areas." June 25. https://www.fcc.gov/reports-research/maps/auction-904-updated-jun20-eligible-areas/.

Federal Communications Commission. 2020c. "Keep Americans Connected." https://www.fcc.gov/keep-americans-connected.

Federal Communications Commission. 2020d. "More about Census Blocks." https://transition.fcc.gov/form477/Geo/more_about_census_blocks.pdf.

Federal Communications Commission. 2020e. *Notice of Proposed Rulemaking and Order* (FCC 20-52). Washington, DC: Federal Communications Commission. https://www.fcc.gov/document/fcc-proposes-5g-fund-rural-america-0.

Federal Communications Commission. 2020f. *Report and Order* (FCC 20-5, WC Docket No. 19-126). Washington, DC: Federal Communications Commission. https://www.fcc.gov/ecfs/filing/02070806418528.

Federal Communications Commission. 2020g. "Rural Digital Opportunity Fund Phase I Auction Complete Applications." https://docs.fcc.gov/public/attachments /DA-20-960A2.pdf.

Federal Communications Commission. 2020h. "Wireline Competition Bureau Confirms That Community Use of E-Rate Supported WI-FI Networks Is Permitted during School and Library Closures Due to COVID-19 Pandemic." https://docs.fcc .gov/public/attachments/DA-20-324A1.pdf.

Ferguson, Scott. 2018. "John Deere Bets the Farm on AI, IoT." *LightReading*, March 12. https://www.lightreading.com/enterprise-cloud/machine-learning-and-ai/john -deere-bets-the-farm-on-ai-iot/a/d-id/741284.

Fischer, Claude S. 1987a. "The Revolution in Rural Telephony, 1900–1920." *Journal of Social History* 21, no. 1: 5–26. https://doi.org/10.1353/jsh/21.1.5.

Fischer, Claude S. 1987b. "Technology's Retreat: The Decline of Rural Telephony in the United States, 1920–1940." *Social Science History* 11, no. 3: 295–327. https://doi .org/10.2307/1171172.

Fischer, Claude S. 1992. *America Calling: A Social History of the Telephone to 1940.* Berkeley: University of California Press.

Fishbane, Lara and Adie Tomer. 2020. "As Classes Move Online during COVID-19, What Are Disconnected Students to Do?" Brookings Institute. https://www .brookings.edu/blog/the-avenue/2020/03/20/as-classes-move-online-during-covid -19-what-are-disconnected-students-to-do/.

Fletcher, Bevin. 2020. "AT&T Expedites Cell Tower Deployment to Support COVID-19 Medical Facility." *FierceWireless*, April 10. https://www.fiercewireless.com/wireless /at-t-expedites-cell-tower-deployment-to-support-covid-19-medical-facility.

Food, Conservation, and Energy Act of 2008. H.R. 2419. 110th Cong. Pub. L. No. 110-234. https://www.congress.gov/bill/110th-congress/house-bill/2419/text.

Ford, George S. 2016. *The Impact of Government-Owned Broadband Networks on Private Investment and Consumer Welfare.* Washington, DC: State Government Leadership Foundation. https://sglf.org/wp-content/uploads/sites/2/2016/04/SGLF-Muni -Broadband-Paper.pdf.

Ford, George S., and R. Alan Seals. 2019. *The Rewards of Municipal Broadband: An Econometric Analysis of the Labor Market.* Washington, DC: Phoenix Center. https:// www.phoenix-center.org/pcpp/PCPP54Final.pdf.

Fortney, John. C., James F. Burgess, Hayden B. Bosworth, Brenda M. Booth, and Peter. J. Kaboli. 2011. "A Re-Conceptualization of Access for 21st Century Healthcare." *Journal of General Internal Medicine* 26, no. 2: 639–647. https://doi.org/10.1007/s11606-011 -1806-6.

Freedman, Des. 2010. "Media Policy Silences: The Hidden Face of Communications Decision Making." *International Journal of Press/Politics* 15, no. 3: 344–361. https://doi .org/10.1177/1940161210368292.

Freedman, Des. 2014. *The Contradictions of Media Power*. London: Bloomsbury.

Freedman, Des. 2008. *The Politics of Media Policy*. Cambridge: Polity.

Friedland, William. H. 2002. "Agriculture and Rurality: Beginning the 'Final Separation'?." *Rural Sociology* 67, no. 3: 350–371. https://doi.org/10.1111/j.1549-0831.2002 .tb00108.x.

Frischmann, Brett, and Barbara van Schewick. 2007. "Network Neutrality and the Economics of an Information Superhighway: A Reply to Professor Yoo." *Jurimetrics*, 47, no. 4: 383–428.

Frontier. 2020. "Re: Connect America Fund, WC Docket No. 10-90." https://ecfsapi.fcc .gov/file/10115075584633/FTR%20YE%202019%20CAF%20Letter%201.15.20.pdf.

Gagliordi, Natalie. 2018. "How 5G Will Impact the Future of Farming and John Deere's Digital Transformation." *ZDNet*, February 2. https://www.zdnet.com/article/how-5g -will-impact-the-future-of-farming-and-john-deeres-digital-transformation.

Gallardo, Roberto. 2016. *Responsive Countryside: The Digital Age and Rural Communities*. Starkville: Mississippi State University Extension Service.

Gallardo, Roberto, and Cheyanne Geideman. 2019. "Digital Distress: What Is It and Who Does It Affect? Part 1." *Medium*, February 19. https://medium.com/design-and -tech-co/digital-distress-what-is-it-and-who-does-it-affect-part-1-e1214f3f209b.

Gallardo, Roberto, and Brian Whitacre. 2019. *A Look at Broadband Access, Providers and Technology*. Bowling Green, OH: Center for Regional Development. https:// pcrd.purdue.edu/files/media/008-A-Look-at-Broadband-Access-Providers-and -Technology.pdf.

Gault, Matthew. 2019. "Lawmaker Kills Repair Bill Because 'Cellphones Are Throwaways.'" *Vice*, October 25. https://www.vice.com/en_us/article/43kp8j/lawmaker-kills -repair-bill-because-cellphones-are-throwaways.

George, Cherian. 2017. *Communicating with Power*. New York: Peter Lang.

Gibson, Jane. 2019. "Automating Agriculture: Precision Technologies, Agbots, and the Fourth Industrial Revolution." In *In Defense of Farmers: The Future of Agriculture in the Shadow of Corporate Power*, edited by Jane Gibson and Sara Alexander, 135–174. Lincoln: University of Nebraska Press.

Gilroy, Angela, and Lennard Kruger. 2012. *Rural Broadband: The Roles of the Rural Utilities Service and the Universal Service Fund* (R42524). Washington, DC: Congressional Research Service. https://fas.org/sgp/crs/misc/R42524.pdf.

Glaser, Barney, and Anselm Strauss. 1967. *The Discovery of Grounded Theory: Strategies for Qualitative Research*. Piscataway, NJ: Aldine Transaction.

Glass, Victor, and Timothy Tardiff. 2019. "The Federal Communications Commission's Rural Infrastructure Auction: What Is Hidden in the Weeds? *Telecommunications Policy* 43, no. 8. https://doi.org/10.1016/j.telpol.2019.04.005.

Goldman, David. 2018. "What Is 5G?" *CNNMoney*, January 29. https://money.cnn.com/2018/01/29/technology/what-is-5g/index.html.

Gonzalez, Lisa. 2018. "Frontier Under Investigation, Minnesota PUC Schedules Fall Public Hearings." *Community Networks*, June 7. https://muninetworks.org/content/frontier-under-investigation-minnesota-puc-schedules-fall-public-hearings.

Gonzalez, Lisa. 2019. "Illinois Will Invest $420 Million in Broadband as Part of Massive Infrastructure Plan." *Community Networks*, August 26. https://muninetworks.org/content/illinois-will-invest-massive-infrastructure-plan.

Government Accountability Office. 2012. *Recovery Act: Broadband Programs Are Ongoing, and Agencies' Efforts Would Benefit from Improved Data Quality* (GAO-12-937). Washington, DC: Government Accountability Office. http://www.gao.gov/products/GAO-12-937.

Government Accountability Office. 2014. *Recovery Act: USDA Should Include Broadband Program's Impact in Annual Performance Reports* (GAO-14-511). Washington, DC: Government Accountability Office. https://www.gao.gov/products/GAO-14-511.

Government Accountability Office. 2018. *Broadband Internet: FCC's Data Overstate Access on Tribal Lands* (GAO-18-630). Washington, DC: Government Accountability Office. https://www.gao.gov/assets/700/694386.pdf.

Governor's Task Force on Broadband (Minnesota). 2012. *Annual Report and Broadband Plan*. Saint Paul: Governor's Task Force on Broadband. https://www.leg.state.mn.us/docs/2012/other/121294.pdf.

Governor's Task Force on Broadband (Minnesota). 2018. *Annual Report*. Saint Paul: Governor's Task Force on Broadband. https://www.leg.state.mn.us/docs/2018/other/180963.pdf.

Greene, Tim. 2016. "John Deere Is Plowing IoT into Its Farm Equipment." *NetworkWorld*, May 17. https://www.networkworld.com/article/3071340/john-deere-is-plowing-iot-into-its-farm-equipment.html.

Grubesic, Tony H. 2003. "Inequities in the Broadband Revolution." *Annals of Regional Science* 37, no. 2: 263–290. https://doi.org/10.1007/s001680300123.

Grubesic, Tony H. 2006. "A Spatial Taxonomy of Broadband Regions in the United States." *Information Economics & Policy 18*, no. 4: 423–448. https://doi.org/10.1016/j.infoecopol.2006.05.001.

Grubesic, Tony H. 2008. "The Spatial Distribution of Broadband Providers in the United States: 1999–2004." *Telecommunications Policy* 32, nos. 3–4: 212–233. https://doi.org/10.1016/j.telpol.2008.01.001.

Grubesic, Tony H. 2010. "Efficiency in Broadband Service Provision: A Spatial Analysis." *Telecommunications Policy* 34, no. 3: 117–131. https://doi.org/10.1016/j.telpol.2009.11.017.

Grubesic, Tony H., and Elizabeth A. Mack. 2017. *Broadband Telecommunications and Regional Development.* New York: Routledge.

Grush, Loren. 2020. "With Latest Starlink Launch, SpaceX Touts 100Mbps Download Speeds snd 'Space Lasers.'" *The Verge*, September 3. https://www.theverge.com/2020/9/3/21419841/spacex-starlink-internet-satellite-constellation-download-speeds-space-lasers.

Guterres, Antonio. 2020. "Digital Divide 'A Matter of Life and Death' amid COVID-19 Crisis." United Nations. https://www.un.org/press/en/2020/sgsm20118.doc.htm.

Habermas, Jürgen. 1985. "A Philosophico-Political Profile." *New Left Review* 151: 75–105.

Hachman, Mark. 2020. "Data Caps on Comcast, T-Mobile Return on July 1." *PCWorld*, July 1. https://www.pcworld.com/article/3564590/data-caps-on-comcast-t-mobile-return-on-july-1.html.

Hadwiger, Don F., and Clay Cochran. 1984. "Rural Telephones in the United States." *Agricultural History* 58, no. 3: 221–238.

Halfacree, Keith H. 1993. "Locality and Social Representation: Space, Discourse and Alternative Definitions of the Rural." *Journal of Rural Studies* 9, no. 1: 23–37. https://doi.org/10.1016/0743-0167(93)90003-3.

Hambly, Helen, and Mamun Chowdury. 2018. "A Gap Analysis of Broadband Connectivity and Precision Agriculture Adoption in Southwestern Ontario, Canada." In *Proceedings of the 14th International Conference on Precision Agriculture, Montreal, Quebec, Canada, June 24–27, 2018*, 1–15. Monticello, IL: International Society of Precision Agriculture. https://www.ispag.org/proceedings/?action=download&item=5814.

Hammersley, Martyn, and Paul Atkinson. 1995. *Ethnography: Principles in Practice.* 2nd ed. London: Routledge.

Handley, Lucy. 2019. "Amazon Beats Apple and Google to Become the World's Most Valuable Brand." *CNBC*, June 11. https://www.cnbc.com/2019/06/11/amazon-beats-apple-and-google-to-become-the-worlds-most-valuable-brand.html.

Harley, John Brian. 1988. "Maps, Knowledge Power." In *The Iconography of Landscape*, edited by Denis Cosgrove, Stephen Daniels, and Alan R. H. Baker, 277–312. Cambridge: Cambridge University Press.

Harvey, David. 1991. *The Condition of Postmodernity: An Enquiry into the Origins of Cultural Change*. Malden, MA: Wiley-Blackwell.

Harvey, Donna. 2019. "Perceptions of and Policy Making around Aging in Rural America." *Generations* 43, no. 2: 66–70.

Health and Economic Recovery Omnibus Emergency Solutions Act of 2020. H.R. 6800 116th Cong. https://www.congress.gov/bill/116th-congress/house-bill/6800/actions.

Heaven, Will. 2020. "Why the Coronavirus Lockdown Is Making the Internet Stronger Than Ever." *MIT Technology Review*, April 7. https://www.technologyreview.com/2020/04/07/998552/why-the-coronavirus-lockdown-is-making-the-internet-better-than-ever/.

Hendrickson, Clara, Mark Muro, and William Galston. 2015. *Countering the Geography of Discontent: Strategies for Left-Behind Places*. Washington, DC: Brookings Institution. https://www.brookings.edu/research/countering-the-geography-of-discontent-strategies-for-left-behind-places.

Hendrickson, Mary, Philip H. Howard, and Douglas Constance. 2019. "Power, Food and Agriculture: Implications for Farmers, Consumers and Communities." In *In Defense of Farmers: The Future of Agriculture in the Shadow of Corporate Power*, edited by Jane Gibson and Sara Alexander, 13–63. Lincoln: University of Nebraska Press.

Herszenhorn, David M. 2009. "Internet Money in Fiscal Plan: Wise or Waste?." *New York Times*, February 2. https://www.nytimes.com/2009/02/03/us/politics/03broadband.html.

Herzog, Christian, and Christopher Ali. 2015. "Elite Interviewing in Media and Communications Policy Research." *International Journal of Media & Cultural Politics* 11, no. 1: 37–58.

Herzog, Christian, Christian Handke, and Erik Hitters. "Analyzing Talk and Text II: Thematic Analysis." In *The Palgrave Handbook of Methods for Media Policy Research*, edited by Hilde Van den Bulck, Manuel Puppis, Karen Donders, and Leo Van Audenhove, 385–402. Cham, Switzerland: Palgrave Macmillan.

Hesse-Biber, Sharlene J., and Patricia L. Leavy. 2010. *The Practice of Qualitative Research*. 2nd ed. London: SAGE.

Hest, David. 2010. "Telematics 2.0." *Farm Progress*, September 29. https://www.farmprogress.com/precision-farming/telematics-20.

Higgins, Jessie. 2020. "COVID Has Made Rural Schools Suddenly Responsible for Getting Internet to Kids in Remote, Unserved Areas." *Charlottesville Tomorrow*, August 27. https://www.cvilletomorrow.org/articles/covid-has-made-rural-schools -suddenly-responsible-for-getting-internet-to-kids-in-remote-unserved-areas.

Hill, Kelly. 2020. "FCC Touts Use of 5.9 GHz by WISPs to Bolster Rural and Suburban Broadband." *RCRWireless*, May 5. https://www.rcrwireless.com/20200505/spectrum /fcc-touts-use-of-5-9-ghz-by-wisps-to-bolster-rural-and-suburban-broadband.

Hindman, Douglas. B. 2000. "The Rural-Urban Digital Divide." *Journalism & Mass Communication Quarterly* 77, no. 3: 549–560. https://doi.org/10.1177/107769900 007700306.

Hirsh, Richard. 2018. "Shedding New Light on Rural Electrification: The Neglected Story of Successful Efforts to Power Up Farms in the 1920s and 1930s." *Agricultural History* 92, no. 3: 296–327. https://doi.org/10.3098/ah.2018.092.3.296.

Hite, James. 1997. "The Thunen Model and the New Economic Geography as a Paradigm for Rural Development Policy." *Review of Agricultural Economics* 19, no. 2: 230–240. https://doi.org/10.2307/1349738.

Hoffman, Donna L., and Thomas P. Novak. 1998. "Bridging the Racial Divide on the Internet." *Science* 280, no. 5362: 390–391. https://doi.org/10.1126/science.280.5362.390.

Hoffman, Donna L., Thomas P. Novak, and Ann Schlosser. 2000. "The Evolution of the Digital Divide: How Gaps in Internet Access May Impact Electronic Commerce." *Journal of Computer-Mediated Communication* 5, no. 3. https://doi.org/10.1111/j.1083 -6101.2000.tb00341.x.

Hofmokl, Justyna. 2010. "The Internet Commons: Towards an Eclectic Theoretical Framework." *International Journal of the Commons* 4, no. 1: 226–250.

Holma, Harri, and Antti Toskala. 2011. *LTE for UMTS: Evolution to LTE-Advanced.* Chichester, UK: John Wiley & Sons.

Holpuch, Amanda. 2020. "US's Digital Divide 'Is Going to Kill People' as COVID-19 Exposes Inequalities." *Guardian*, April 13. https://www.theguardian.com/world /2020/apr/13/coronavirus-covid-19-exposes-cracks-us-digital-divide.

Horrigan, John. 2020. *Measuring the Gap: What's the Right Approach to Exploring Why Some Americans do not Subscribe to Broadband?* Washington, DC: National Digital Inclusion Alliance. https://www.digitalinclusion.org/measuring-the-gap/.

Horrigan, John, and Maeve Duggan. 2015. "Barriers to Broadband Adoption: Cost Is Now a Substantial Challenge for Many Non-Users." Pew Research Center. December 21. https://www.pewresearch.org/internet/2015/12/21/3-barriers-to-broadband -adoption-cost-is-now-a-substantial-challenge-for-many-non-users/.

Horwitz, Robert. 1989. *The Irony of Regulatory Reform: The Deregulation of American Telecommunications.* Oxford: Oxford University Press.

Horwitz, Jeremy. 2019. "The Definitive Guide to 5G Low, Mid, and High Band Speeds." *VentureBeat*, December 10. https://venturebeat.com/2019/12/10/the-definitive-guide -to-5g-low-mid-and-high-band-speeds/.

House of Commons Library. 2020. "Broadband." UK Parliament. October 1. https:// commonslibrary.parliament.uk/broadband-faqs/.

Howard, Phillip H. 2016. *Concentration and Power in the Food System: Who Controls What We Eat?* London: Bloomsbury Academic.

Howarth, David. 2000. *Discourse*. Buckingham, UK: Open University Press.

Howell, Bronwyn, and Petrus Potgieter. 2020. "Politics, Policy and Fixed-Line Tele-communications Provision: Insights from Australia." *Telecommunications Policy* 44. https://doi.org/10.1016/j.telpol.2020.101999.

Howell, Catherine, and Daniel West. 2016. "The Internet as a Human Right." *TechTank*, November 7. https://www.brookings.edu/blog/techtank/2016/11/07/the -internet-as-a-human-right/.

Hu, Jane C. 2018. "Everything You Need to Know about 5G." *Quartz*, October 31. https://qz.com/1442559/everything-you-need-to-know-about-5g.

Hu, Tung-Hui. 2016. *A Prehistory of the Cloud*. Cambridge, MA: MIT Press.

Hughes Network Systems. 2009. "Ex Parte Notice (GN Docket No. 09-29)." March 24. https://www.fcc.gov/ecfs/filing/5515350206.

Hughes Network Systems. 2019. "CAF-II Metrics Order Petitions for Reconsideration." April 5. https://ecfsapi.fcc.gov/file/1040932062890/April%205%20HNS%20Ex%20 Parte.pdf.

Ilbery, Brian 1998. "Dimensions of Rural Change." In *The Geography of Rural Change*, edited by Brian Ilbery, 1–10. New York: Routledge.

Institute for Local Self-Reliance. 2019. "Cooperatives Build Community Networks." Community Networks. https://muninetworks.org/content/rural-cooperatives-page.

International Cooperative Alliance. n.d. "Cooperative Identity, Values & Principles." https://www.ica.coop/en/cooperatives/cooperative-identity#cooperative-principles.

International Telecommunications Union. 2015. "Global Broadband Growth Slows Sharply: 4 Billion Still Offline." https://www.itu.int/net/pressoffice/press_releases /2015/35.aspx.

International Telecommunications Union. 2019. *The State of Broadband: Broadband as a Foundation for Sustainable Development*. Geneva: International Telecommuni-cations Union and UNESCO. https://www.itu.int/dms_pub/itu-s/opb/pol/S-POL -BROADBAND.20-2019-PDF-E.pdf.

ITTA—The Voice of America's Broadband Providers. 2018. "Comments of ITTA—The Voice of America's Broadband Providers" (RUS-18-Telecom-004). Regulations .gov. https://www.regulations.gov/document?D=RUS-18-TELECOM-0004-0213.

Janzen, Todd. 2017. "Why Dealers Should Pay Attention to the Fine Print." *Precision Farming Dealer*, February 4. https://www.precisionfarmingdealer.com/articles/2618 -why-dealers-should-pay-attention-to-the-fine-print.

Jasmontaite, Lina, and Paul de Hert. 2019. "Access to the Internet in the EU: A Policy Priority, a Fundamental, a Human Right, or a Concern for eGovernment?" In *Research Handbook on Human Rights and Digital Technology: Global Politics, Law and International Relations*, edited by Ben Wagner, Matthias C. Ketteman, and Kilian Vieth, 157–179. Cheltenham, UK: Edward Elgar.

John, Richard. 2010. *Network Nation:* Inventing American Telecommunications. Cambridge, MA: Harvard University Press.

Johnson, Kenneth 2014. "Demographic Trends in Nonmetropolitan America." In *Rural America in a Globalizing World*, edited by Conner Bailey, Leif Jensen, and Elizabeth Ransom, 311–329. Morgantown: West Virginia University Press.

Johnson, Renee, and Jim Monke. 2018. *What Is the Farm Bill?* (RS22131). Washington, DC: Congressional Research Service. https://fas.org/sgp/crs/misc/RS22131 .pdf.

Johnston, Louis D. 2016. "No, Broadband and Garbage Collection Aren't Public Goods. Here's Why They Might Require Regulation Anyway." *MinnPost*, August 5. https://www.minnpost.com/macro-micro-minnesota/2016/08/no-broadband-and -garbage-collection-aren-t-public-goods-here-s-why-the/.

Jones, Sarah. 2018. "Telling Rural People to Move Won't Solve the Problem." *New Republic*, January 23. https://newrepublic.com/article/146713/telling-rural-people -move-wont-solve-poverty.

Kahan, John. 2019. "It's Time for a New Approach for Mapping Broadband Data to Better Serve Americans." *Microsoft on the Issues*, April 8. https://blogs.microsoft.com /on-the-issues/2019/04/08/its-time-for-a-new-approach-for-mapping-broadband -data-to-better-serve-americans.

Kandilov, Amy M. G., Ivan T. Kandilov, Xiangping Liu, and Mitch Renkow. 2017. "The Impact of Broadband on U.S. Agriculture: An Evaluation of the USDA Broadband Loan Program." *Applied Economic Perspectives and Policy* 39, no. 4: 635–661. https://doi.org/10.1093/aepp/ppx022.

Kandilov, Ivan. T., and Mitch Renkow. 2010. "Infrastructure Investment and Rural Economic Development: An Evaluation of USDA's Broadband Loan Program." *Growth & Change* 41, no. 2: 165–191. https://doi.org/10.1111/j.1468-2257.2010.00524.x.

Kang, Cecilia. 2017. "How to Give Rural America Broadband? Look to the Early 1900s." *New York Times*, December 21. https://www.nytimes.com/2016/08/08/technology /how-to-give-rural-america-broadband-look-to-the-early-1900s.html.

Kang, Cecilia. 2020. "Parking Lots Have Become a Digital Lifeline." *New York Times*, May 5. https://www.nytimes.com/2020/05/05/technology/parking-lots-wifi-coronavirus .html.

Kaur, Harmeet. 2020. "Why Rural Americans Are Having a Hard Time Working from Home." *CNN*, April 29. https://www.cnn.com/2020/04/29/us/rural-broadband -access-coronavirus-trnd/index.html.

Keillor, Steven J. 2000. *Cooperative Commonwealth*. St. Paul: Minnesota Historical Society Press.

Khazan, Olga. 2020. "America's Terrible Internet Is Making Quarantine Worse." *The Atlantic*, August 17. https://www.theatlantic.com/technology/archive/2020/08 /virtual-learning-when-you-dont-have-internet/615322/.

Kim, Younjun, and Peter F. Orazem. 2017. "Broadband Internet and New Firm Location Decisions in Rural Areas." *American Journal of Agricultural Economics* 99, no. 1: 285–302. https://doi.org/10.1093/ajae/aaw082.

Kincheloe, Joe, and Peter McLaren. 1994. "Rethinking Critical Theory and Qualitative Research." In *The SAGE Handbook of Qualitative Research*, edited by Norm Denzin and Yvonna Lincoln, 138–157. Thousand Oaks: SAGE.

Kinney, Sean. 2019. "What's the Outlook for 5G Spectrum Harmonization in Europe?" *RCRWireless*, August 23. https://www.rcrwireless.com/20190823/5g/5g-spectrum -harmonization-europe.

Kitchin, Rob. 2014. *The Data Revolution*. Thousand Oaks, CA: SAGE.

Klein, Dorothea. 2013. *Technologies of Choice?: ICTs, Development, and the Capabilities Approach*. Cambridge, MA: MIT Press.

Kline, Ronald R. 2000. *Consumers in the Country: Technology and Social Change in Rural America*. Baltimore: Johns Hopkins University Press.

Koebler, Jason. 2017. "Why American Farmers Are Hacking Their Tractors with Ukrainian Firmware." *Motherboard*, March 21. https://www.vice.com/en_us/article/xykkkd /why-american-farmers-are-hacking-their-tractors-with-ukrainian-firmware.

Koebler, Jason. 2018a. "Tractor-Hacking Farmers Are Leading a Revolt against Big Tech's Repair Monopolies." *Motherboard*, February 14. https://www.vice.com/en_us /article/kzp7ny/tractor-hacking-right-to-repair.

Koebler, Jason. 2018b. "Tractor Hacking: Watch Our Documentary about Farmers Fighting for the Right to Repair." *Motherboard*, February 15. https://www.vice

.com/en_us/article/pamkqn/watch-tractor-hacking-john-deere-right-to-repair
-documentary.

Koeppel, Dan. 2019. "Moving to the Woods Killed My Internet. Here's What I Did
about It." *Wirecutter*, September 11. https://www.nytimes.com/wirecutter/blog
/moving-to-the-wilderness-killed-my-internet/.

Koppell, Jonathan. G. 2005. "Pathologies of Accountability: ICANN and the Challenge of 'Multiple Accountabilities Disorder.'" *Public Administration Review* 65, no. 1:
94–108. https://doi.org/10.1111/j.1540-6210.2005.00434.x.

Kruger, Lennard. 2018. *Broadband Loan and Grant Programs in the USDA's Rural Utilities Service* (RL33816). Washington, DC: Congressional Research Service. https://www
.everycrsreport.com/files/20181016_RL33816_76629ba2fb086f856e1d10a148ff0cf4ac
a53cbd.html.

Kruger, Lennard. 2019. *Broadband Loan and Grant Programs in the USDA's Rural Utilities Service* (RL33816). Washington, DC: Congressional Research Service. https://fas
.org/sgp/crs/misc/RL33816.pdf.

Kruger, Lennard, and Angela Gilroy. 2016. *Municipal Broadband: Background and
Policy Debate* (R44080). Washington, DC: Congressional Research Service. https://
www.fas.org/sgp/crs/misc/R44080.pdf.

Kuttner, Hans. 2016. *The Economic Impact of Rural Broadband*. Washington, DC:
Hudson Institute. https://s3.amazonaws.com/media.hudson.org/files/publications/2
0160419KuttnerTheEconomicImpactofRuralBroadband.pdf.

Landgraf, Steven. 2020. "Entry Threats from Municipal Broadband Internet and
Impacts on Private Provider Quality." *Information Economics and Policy* 52. https://
doi.org/10.1016/j.infoecopol.2020.100878.

Land O'Lakes. 2020. "American Connection Project Broadband Coalition." https://
www.landolakesinc.com/Press/News/American-Connection-Project-Broadband
-Coalition.

Lanner. 2018. "5G and Smart Farming IoT—Promise of Making the World Green Again."
July 30. https://www.lanner-america.com/blog/smart-farming-iot-5g-agriculture/.

Lasswell, Harold. 2003. "The Policy Orientation." In *Communication Researchers and
Policy-Making*, edited by Sandra Braman, 85–104. Cambridge, MA: MIT Press.

Latzke, Paul. 1906. *A Fight with an Octopus: Being the Story of a Great Contest That
Was Won against Tremendous Odds, as Printed Originally in* Success Magazine. Chicago:
Telephony Pub. Co.

Lawless, Paul, and Tony Gore. 1999. "Urban Regeneration and Transport Investment: A Case Study of Sheffield 1992–96." *Urban Studies* 36, no. 3: 527–545. https://
doi.org/10.1080/0042098993510.

LeDuc, Don. 2003. "Transforming Principles into Policy." In *Communication Research-ers and Policy-Making*, edited by Sandra Braman, 167–176. Cambridge, MA: MIT Press.

Lefebvre, Henri. 1995. *Writings on Cities*. Malden, MA: Wiley-Blackwell.

Lefebvre, Henri. 2016a. *Marxist Thought and the City*. Translated by R. Bononno. Minneapolis: University of Minnesota Press.

Lefebvre, Henri. 2016b. "The Theory of Ground Rent and Rural Sociology." *Antipode* 48, no. 1: 67–73. https://doi.org/10.1111/anti.12172.

Lefebvre, Henri, and Mario Gaviria. 1973. *Du rural à L'urbain*. Sankt Augustin, Germany: Éditions Anthropos.

Legere, John. 2019. "New T-Mobile: Bridging the Digital Divide . . . for GOOD." T-Mobile. April 17. https://www.t-mobile.com/news/new-t-mobile-bridging-digital -divide.

Levin, Blair, and Larry Downes. 2019. "Cities, Not Rural Areas, Are the Real Internet Deserts."

Washington Post, September 13. https://www.washingtonpost.com/technology/2019 /09/13/cities-not-rural-areas-are-real-internet-deserts/.

Levitz, Jennifer, and Valerie Bauerlein. 2017. "Rural America Is Stranded in the Dial-Up Age." *Wall Street Journal*, June 15. https://www.wsj.com/articles/rural-america-is -stranded-in-the-dial-up-age-1497535841.

Lewis, Justin. 1991. *The Ideological Octopus: An Exploration of Television and Its Audience*. New York: Routledge.

Locke, Terry. 2004. *Critical Discourse Analysis*. New York: Continuum.

Lowrey, Annie. 2017. "The Great Recession Is Still with Us." *The Atlantic*, December 1. https://www.theatlantic.com/business/archive/2017/12/great-recession-still-with-us /547268.

Lukes, Steven. 2005. *Power: A Radical View*. 2nd ed. New York: Palgrave Macmillan.

MacDougall, Robert. 2014. *The People's Network: The Political Economy of the Telephone in the Gilded Age*. Philadelphia: University of Pennsylvania Press.

Malecki, Edward. J. 2003. "Digital Development in Rural Areas: Potentials and Pitfalls." *Journal of Rural Studies* 19, no. 2: 201–214. https://doi.org/10.1016/S0743 -0167(02)00068-2.

Malecki, Edward J., and Bruno Moriset. 2003. *The Digital Economy: Business Organization, Production Processes, and Regional Developments*. New York: Routledge.

Mansell, Robin. 2004. "Political Economy, Power and New Media." *New Media & Society* 6, no. 1: 96–105. https://doi.org/10.1177/1461444804039910.

Mansell, Robin. 2012. *Imagining the Internet: Communication, Innovation, and Governance*. Oxford: Oxford University Press.

Marcattilio-McCracken, Ry. 2020. "Champaign, Illinois Brings Fixed Wireless to Students in Mobile Home Park." Community Networks. September 15. https://muninetworks.org/content/champaign-illinois-brings-fixed-wireless-students-mobile-home-park.

Marek, Sue. 2018. "John Deere's 5G Aspirations Include Streaming Video." *SDxCentral*, March 21. https://www.sdxcentral.com/articles/news/john-deeres-5g-aspirations-include-streaming-video-autonomous-tractors/2018/03.

Mariscal, Judith. 2020. "The Case of the Wholesale Mobile Network in Mexico: Red Compartida." In *Frequencies:* International Spectrum Policy, edited by Gregory Taylor and Catherine Middleton, 116–137. Montreal: McGill-Queens University Press.

Mark, Tyler B., Terry W. Griffin, and Brian E. Whitacre. 2016. "The Role of Wireless Broadband Connectivity on 'Big Data' and the Agricultural Industry in the United States and Australia." *International Food and Agribusiness Management Review* 19, no. A: 43–56. https://www.ifama.org/resources/Documents/v19ia/220150113.pdf.

Marsh, Charles, Peter Slade, and Sarah Azaransky. 2017. *Lived Theology: New Perspectives on Method, Style, and Pedagogy*. Oxford: Oxford University Press.

Marsh, Joan. 2020. "We Need to Fundamentally Rethink How USF Programs Are Funded." AT&T Public Policy. July 21. https://www.attpublicpolicy.com/universal-service/we-need-to-fundamentally-rethink-how-usf-programs-are-funded/.

Marvin, Carolyn. 1988. *When Old Technologies Were New: Thinking about Electric Communication in the Late Nineteenth Century*. Oxford: Oxford University Press.

Marx, Karl. 1888. *Theses on Feuerbach*. In *Marx/Engels Selected Works*, translated by W. Lough, 13–15. Moscow: Progress Publishers. https://www.marxists.org/archive/marx/works/1845/theses/theses.htm.

Marx, Karl. [1939] 1993. *Grundrisse: Foundations of the Critique of Political Economy*. Translated by M. Nicolaus. New York: Penguin Classics.

Marx, Karl. [1867] 2004. *Capital: A Critique of Political Economy*. London: Penguin.

Matheny, Paul D. 2012. *Contextual Theology: The Drama of Our Times*. Havertown, PA: Casemate Publishers.

Matisse, Nathan. 2020. "In the COVID-19 Era, the Wheels on the Bus Increasingly Bring Wi-Fi." *Ars Technica*, April 15. https://arstechnica.com/information-technology/2020/04/in-the-covid-19-era-the-wheels-on-the-bus-increasingly-bring-wi-fi/.

Maxwell, Winston, and Marc Bourreau. 2014. "Technological Neutrality in Internet, Telecoms and Data Protection Regulation." *Communication Technology Law Review* 1:

1–4. https://www.hoganlovells.com/~/media/hogan-lovells/pdf/publication/201521ctlr
issue1maxwell_pdf.pdf.

McChesney, Robert W. 1995. *Telecommunications, Mass Media, and Democracy: The
Battle for the Control of U.S. Broadcasting, 1928–1935*. Oxford: Oxford University Press.

McGill, Margaret Harding. 2020. "Pandemic Sparks a Run on Hotspot Devices for
Students." *Axios*, April 1. https://www.axios.com/hotspots-wifi-students-coronavirus
-1a08db65-236b-43e8-a5e0-17429a07942a.html.

McIntosh, Matt. 2018. "The Legal Mess of Farm Data Ownership." *Farmtario*, September 24. https://farmtario.com/machinery/the-legal-mess-of-farm-data-ownership.

McKenzie, Lindsay. 2020. "Limited Broadband Access Not Just Rural Issue." *Inside
Higher Ed*, August 13. https://www.insidehighered.com/quicktakes/2020/08/13/limited
-broadband-access-not-just-rural-issue.

McKinion, James M. 2010. "Role of Telecommunications in Precision Agriculture."
In *E-Agriculture and E-Government for Global Policy Development: Implications and Future
Directions*, edited by Blessing Maumbe, 252–266. Hershey, PA: Information Science
Publishing.

McKinnon, John, and Ryan Tracy. 2020. "Pandemic Builds Momentum for Broadband
Infrastructure Upgrade." *Wall Street Journal*, April 23. https://www.wsj.com/articles
/pandemic-builds-momentum-for-broadband-infrastructure-upgrade-11587461400.

McLuhan, Marhsall. 1964. *Understanding Media: The Extensions of Man*. New York:
McGraw-Hill.

Medema, Steven. 2009. *The Hesitant Hand: Taming Self-Interest in the History of Economic Ideas*. Princeton, NJ: Princeton University Press.

Meinrath, Sascha. D. 2019. *Broadband Availability and Access in Rural Pennsylvania*. Harrisburg: Center for Rural Pennsylvania. https://www.rural.palegislature.us/broadband
/Broadband_Availability_and_Access_in_Rural_Pennsylvania_2019_Report.pdf.

Meyrowitz, Johsua. 1985. *No Sense of Place: The Impact of Electronic Media on Social
Behavior*. Oxford: Oxford University Press.

Miller, Alfred. 2019. "Kentuckians Were Promised Internet. What They Got: $1.5B
Information Highway to Nowhere." *Louisville Courier Journal*, May 14. https://www
.courier-journal.com/story/news/investigations/2019/05/08/hristop-wired-internet
-project/3238594002.

Minnesota Department of Employment and Economic Development. 2019.
Office of Broadband Development Annual Report. St. Paul: Minnesota Department of
Employment and Economic Development. https://www.leg.state.mn.us/docs/2019
/mandated/190132.pdf.

Minnesota Department of Employment and Economic Development. 2020. *Minnesota Broadband Infrastructure Plan*. St. Paul: Minnesota Department of Employment and Economic Development. https://mn.gov/deed/assets/state-broadband -plan_tcm1045-380006.pdf.

Minnesota Statute. 2020. Local Improvements, Council Powers. §429.021(19)(ii). https://www.revisor.mn.gov/statutes/cite/429.021

Mitchell, Christopher. 2017. *Correcting Community Fiber Fallacies*. Minneapolis: Institute for Local Self-Reliance. https://muninetworks.org/sites/www.muninetworks.org /files/fiber-fallacy-upenn-yoo.pdf.

Mitchell, Christopher, and Hannah Trostle. 2018. *Profiles of Monopoly: Big Cable and Telecom*. Minneapolis, MN: Institute for Local Self-Reliance. https://ilsr.org/wp -content/uploads/2018/07/profiles-of-monopoly-2018.pdf.

Mitchell, Ronald K., Bradley R. Agle, and Donna J. Wood. 1997. "Toward a Theory of Stakeholder Identification and Salience: Defining the Principle of Who and What Really Counts." *Academy of Management Review* 22, no. 4: 853–886. https://doi.org /10.2307/259247.

M-Lab. 2019. "Louisa." https://viz.measurementlab.net/location/nausvalouisa?isps=AS1 3367x_AS7155_AS11486x.

M-Lab. 2020. "Crozet." https://datastudio.google.com/reporting/1djtGEuqV4Qwrj26 GQTN_xzp3rsMYYcmv/page/YW8NB?s=rzD5rHYkLT4.

Monsanto. 2018. "How Data Is Driving Sustainability for Farmers." June 27. https:// monsanto.com/innovations/data-science/articles/digital-tools-sustainability/.

Moore, Dale. 2018. "Re: Rural Utilities Service e-Connectivity Pilot Program (Docket No. RUS-18-Telecom-0004)." American Farm Bureau Federation. https://www.fb.org /files/broadband-ruspilotprogram9-10-2018.pdf.

Moran, Rachel, and Matthew Bui. 2019. "Race, Ethnicity, and Telecommunications Policy Issues of Access and Representation: Centering Communities of Color and Their Concerns." *Telecommunications Policy* 43, no. 5: 461–473. https://doi.org/10 .1016/j.telpol.2018.12.005.

Morgan, Richard J. n.d. "Rock County Broadband Access Commentary." https://mn .gov/deed/assets/rock-county-sanford_tcm1045-301892.pdf.

Morris, Frank. 2019. "Critics of Relocating USDA Research Agencies Point to Brain Drain." *NPR*, September 10. https://www.npr.org/2019/09/10/759053717/critics-of -relocating-usda-research-agencies-point-to-brain-drain.

Mosco, Vincent. 2005. *The Digital Sublime: Myth, Power, and Cyberspace*. Cambridge, MA: MIT Press.

Mosco, Vincent. 2009. *The Political Economy of Communication.* 2nd ed. Thousand Oaks, CA: SAGE.

Mosco, Vincent. 2017. *Becoming Digital: Toward a Post-Internet Society.* Bingley, UK: Emerald.

Mossberger, Karen, Caroline J. Tolbert, and Ramona S. McNeal. (2012). *Digital Citizenship: The Internet, Society, and Participation.* Cambridge, MA: MIT Press.

MSBA Advocate. 2017. "Broadband Task Force Observes Benefits of High-Speed Access during Rock County Tour," June 30. https://msbaadvocate.com/2017/06/30/broadband -task-force-observes-benefits-of-high-speed-access-during-rock-county-tour/.

Mueller, Milton. 1997. *Universal Service: Competition, Interconnection and Monopoly in the Making of the American Telephone System.* Washington, DC: American Enterprise Institute.

Napoli, Phillip M. 2001. *Foundations of Communications Policy: Principles and Process in the Regulation of Electronic Media.* Cresskill, NJ: Hampton Press.

National Association of State Utility Consumer Advocates. 2009. "Comments of the National Association of State Utility Consumer Advocates" (GN Docket No. 09-29). https://www.fcc.gov/ecfs/filing/5515348715.

National Digital Inclusion Alliance. 2017. "Definitions." January 18. https://www .digitalinclusion.org/definitions.

National League of Cities. 2016. "Cities 101—Delegation of Power." December 13. https://www.nlc.org/resource/cities-101-delegation-of-power.

National Rural Telecommunications Cooperative. 2009. "Comments of the National Rural Telecommunications Cooperative" (GN Docket No. 09-29, WT Docket No. 07- 293). Federal Communications Commission. https://ecfsapi.fcc.gov/file/6520203466 .pdf.

National Telecommunications and Information Association. 2000. *Falling through the Net: Toward Digital Inclusion.* Washington, DC: National Telecommunications and Information Administration. https://www.ntia.doc.gov/files/ntia/publications /fttn00.pdf.

New Deal Rural Broadband Act of 2017. H.R. 800. 115th Cong. https://www.congress .gov/bill/115th-congress/house-bill/800.

Newton, Harry, and Steve Schoen. 2018. Newton's Telecom Dictionary, 31st ed. New York: Telecom Publishing.

Next Century Cities. 2019. *Becoming Broadband Ready: A Toolkit for Communities.* Washington, DC: Next Century Cities. https://nextcenturycities.org/wp-content /uploads/Becoming-Broadband-Ready-Toolkit-web.pdf.

Nextlink. 2009. "Re: GN Docket No. 09-29: In the Matter of Rural Broadband Strategy." https://ecfsapi.fcc.gov/file/6520203401.pdf.

Nicholas, Kyle. 2003. "Geo-Policy Barriers and Rural Internet Access: The Regulatory Role in Constructing the Digital Divide." *Information Society 19*, no. 4: 287–295.

Norris, Pippa. 2001. *Digital Divide: Civic Engagement, Information Poverty, and the Internet Worldwide*. Cambridge: Cambridge University Press.

Novak, William J. 2013. "A Revisionist History of Regulatory Capture." In *Preventing Regulatory Capture: Special Interest Influence and How to Limit It*, edited by Daniel Carpenter and David Moss, 25–48. Cambridge: Cambridge University Press.

NTCA—The Rural Broadband Association. 2009. "NTCA Comprehensive Rural Broadband Strategy Comments" (GN Docket No. 09-29). Federal Communications Commission. https://ecfsapi.fcc.gov/file/6520203489.pdf.

NTCA—The Rural Broadband Association. 2015. "Comments to the Broadband Opportunity Council." National Telecommunications and Information Administration. https://www.ntia.doc.gov/federal-register-notice/2015/broadband-opportunity-council-comments.

NTCA—The Rural Broadband Association. 2018. *2018 Broadband Survey Report*. Arlington, VA: NTCA–Rural Broadband Association. https://www.ntca.org/sites/default/files/documents/2018-12/2018%20Broadband%20Survey%20Report_FINAL.pdf.

NTCA—The Rural Broadband Association. 2019. "Reply Comments of NTCA—Rural Broadband Association" (WC Docket Nos. 19-126, 10-90). Federal Communications Commission. https://ecfsapi.fcc.gov/file/1021587925248/10.21.19_NTCA_Reply_Comments_on_RDOF_Dockets_19-126_10-90.pdf.

Null, Eric, and Amir Nasir. 2017. "Christopher Yoo's Municipal Broadband Report Misleads on Viability, Success of Municipal Fiber Networks." New America. July 9. https://www.newamerica.org/oti/blog/christopher-yoos-municipal-broadband-report-misleads-viability-success-municipal-fiber-networks/.

Nye, David E. 1990. *Electrifying America: Social Meanings of a New Technology, 1880–1940*. Cambridge, MA: MIT Press.

O'Donnell, Bob. 2019. "The 5G Landscape, Part 2: Spectrum and Devices." *Forbes*, October 24. https://www.forbes.com/sites/bobodonnell/2019/10/24/the-5g-landscape-part-2-spectrum-and-devices/#3a16159751b8.

"Office of Rural Broadband Act." Congress.gov. February 12. https://www.congress.gov/bill/116th-congress/senate-bill/454/committees.

Ogle, Eric. 2019. "Buy Local (Including Broadband)." *Broadband Communities*, January–February. http://www.bbcmag.com/rural-broadband/buy-local-including-broadband.

O'Hara, B. 2018. "Re: Request for Comments on the Rural Utilities Service (RUS) Notice of Inquiry and Request for Comments on RUS e-Connectivity Pilot Program; 83 Fed. Reg. 35,609 (RUS-18 TELECOM-0004) (July 27, 2018)." National Rural Electric Cooperative Association. https://www.regulations.gov/document?D=RUS-18-TELECOM-0004-0196.

Oldenburg, Ray. 1999. *The Great Good Place: Cafes, Coffee Shops, Bookstores, Bars, Hair Salons, and Other Hangouts at the Heart of a Community*. 3rd ed. Philadelphia, PA: Da Capo Press.

Ookla. 2018. "United States: 2018 Fixed." Speedtest. http://www.speedtest.net /reports/united-states/2018.

Opalka, Alice, Alexis Gable, Tara Nicola, and Jennifer Ash. 2020. "Rural School Districts Can Be Creative in Solving the Internet Connectivity Gap—but They Need Support." Brookings Institute. August 10. https://www.brookings.edu/blog/brown -center-chalkboard/2020/08/10/rural-school-districts-can-be-creative-in-solving-the -internet-connectivity-gap-but-they-need-support/.

Open Ag Data Alliance. n.d. "About Open Ag Data Alliance." http://openag.io/about-us.

Open Ag Data Alliance. 2016. "Principals & Use Cases." http://openag.io/about-us /principals-use-cases.

Organisation for Economic Co-operation and Development. 2018. *Bridging the Rural Digital Divide*. Paris: Organisation for Economic Co-operation and Development. https://www.oecd-ilibrary.org/docserver/852bd3b9-en.pdf?expires=1608073647 &id=id&accname=guest&checksum=240ACE5E72BE35FED25168B9E1755791.

Pagliery, Jose. 2015. "OMG: 2.1 Million People Still Use AOL Dial-Up." *CNNMoney*, May 8. https://money.cnn.com/2015/05/08/technology/aol-dial-up/index.html.

Pai, Ajit. 2017a. Expanded Broadband Can Help Combat Rural Flight. *Wichita Eagle*, September 22. https://www.kansas.com/opinion/opn-columns-blogs/article174579201 .html.

Pai, Ajit. 2017b. "Remarks of FCC Chairman Ajit Pai at the Newseum: 'The Future of Internet Freedom.'" Federal Communications Commission. April 26. https://www .fcc.gov/document/chairman-pai-speech-future-internet-regulation.

Pai, Ajit. 2019a. "Pai Announces Plan to Launch $9 Billion 5G Fund for Rural America." Federal Communications Commission. December 4. https://www.fcc.gov /document/pai-announces-plan-launch-9-billion-5g-fund-rural-america.

Pai, Ajit. 2019b. "Statement of Chairman Ajit Pai re: Rural Digital Opportunity Fund, WC Docket No. 19-126; Connect America Fund, WC Docket No. 10-90." Federal Communications Commission. https://docs.fcc.gov/public/attachments/FCC-19-77A2.pdf.

Parker, Edwin B., Heather E. Hudson, Don A. Dillman, and Andrew D. Roscoe. 1989. *Rural America in the Information Age*. Lanham, MD: University Press of America.

Parker, Kim, Juliana Menasce Horowitz, Anna Brown, Richard Fry, D'Vera Cohn, and Ruth Igielnik. 2018. *What Unites and Divides Urban, Suburban and Rural Communities*. Washington, DC: Pew Research Center. https://www.pewsocialtrends.org/2018/05/22/what-unites-and-divides-urban-suburban-and-rural-communities/.

Peacock, Anne. 2019. *Human Rights and the Digital Divide*. London: Routledge.

Pence, Richard, ed. 1984. *The Next Greatest Thing: 50 Years of Rural Electrification in America*. Washington, DC: National Rural Electric Cooperative Association.

Pennsylvania Farm Bureau. 2014. "Ponder These Nine before You Sign." Camp Hill: Pennsylvania Farm Bureau. https://www.pfb.com/images/issues/Ponder-These-Nine.pdf.

Perdue, Sonny. 2017. *Report to the President of the United States from the Task Force on Agriculture and Rural Prosperity*. Washington, DC: US Department of Agriculture. https://www.usda.gov/sites/default/files/documents/rural-prosperity-report.pdf.

Perdue, Sonny, and Wilbur Ross. 2019. *American Broadband Initiative: Milestones Report*. Washington, DC: US Department of Agriculture. https://www.ntia.doc.gov/files/ntia/publications/american_broadband_initiative_milestones_report.pdf.

Perrin, Andrew. 2019. "Digital Gap between Rural and Nonrural America Persists." *Fact Tank: News in the Numbers*, May 31. https://www.pewresearch.org/fact-tank/2019/05/31/digital-gap-between-rural-and-nonrural-america-persists.

Perzanowski, Aaron, and Jason Schultz. 2016. *The End of Ownership: Personal Property in the Digital Economy*. Cambridge, MA: MIT Press.

Pew Charitable Trusts. 2019. *No One Approach Fits All States in Efforts to Expand Broadband Access*. Washington, DC: Pew Charitable Trusts. https://www.pewtrusts.org/-/media/assets/2019/07/bri_agencies_v1.pdf.

Pew Charitable Trusts. 2020. *How States Are Expanding Broadband Access*. Washington, DC: Pew Charitable Trusts. https://www.pewtrusts.org/-/media/assets/2020/03/broadband_report0320_final.pdf.

Pew Research Center. 2019. "Mobile Fact Sheet." June 12. https://www.pewresearch.org/internet/fact-sheet/mobile/.

Philbrick, Jay. 2020. *State CARES Act Broadband Funding Report*. Augusta: Maine Governor's Economic Recovery Committee.

Philip, Lorna, Caitlin Cottrill, and John Farrington. 2015 "'Two-Speed' Scotland: Patterns and Implications of the Digital Divide in Contemporary Scotland." *Scottish*

Geographical Journal 131, nos. 3–4: 148–170. https://doi.org/10.1080/14702541.2015
.1067327.

Philip, Lorna, Caitlin Cottrill, John Farrington, Fiona Williams, and Fiona Ashmore.
2017. "The Digital Divide: Patterns, Policy and Scenarios for Connecting the 'Final
Few' in Rural Communities across Great Britain." *Journal of Rural Studies* 54: 386–
398. https://doi.org/10.1016/j.jrurstud.2016.12.002.

Phillips, Martin. 1998. "Social Perspectives." In *The Geography of Rural Change*, edited
by Brian W. Ilbery, 31–54. New York: Routledge.

Pickard, Victor. 2013. "Social Democracy or Corporate Libertarianism?: Conflicting
Media Policy Narratives in the Wake of Market Failure." *Communication Theory* 23,
no. 4: 336–355. https://doi.org/10.1111/comt.12021.

Pickard, Victor. 2014a. *America's Battle for Media Democracy: The Triumph of Corporate Libertarianism and the Future of Media Reform*. Cambridge: Cambridge University
Press.

Pickard, Victor. 2014b. "The Great Evasion: Confronting Market Failure in American
Media Policy." *Critical Studies in Media Communication* 31, no. 2: 153–159. https://
doi.org/10.1080/15295036.2014.919404.

Pickard, Victor. 2015. "The Return of the Nervous Liberals: Market Fundamentalism,
Policy Failure, and Recurring Journalism Crises." *Communication Review* 18, no. 2:
82–97. https://doi.org/10.1080/10714421.2015.1031995.

Pickard, Victor. 2019. *Democracy without Journalism?: Confronting the Misinformation
Society*. Oxford: Oxford University Press.

Pickard, Victor, and David Berman. 2019. *After Net Neutrality*. New Haven, CT: Yale
University Press.

Pickard, Victor, and Pawel Popiel. 2018. *The Media Democracy Agenda: The Strategy and
Legacy of FCC Commissioner Michael J. Copps*. Evanston, IL: Benton Institute for Broadband and Society. https://www.benton.org/sites/default/files/Copps_legacy.pdf.

Pool, Ithiel de Sola, ed. 1977. The social impact of the telephone. Cambridge: MIT
Press.

Porter, Jody. 2020. "Schools Resort to Phone and Fax Machine to Restart Classes in
Northern Ontario First Nations." *CBC*, August 27. https://www.cbc.ca/news/canada
/thunder-bay/phone-fax-school-1.5701272.

Porter, Jon. 2019. "Amazon Will Launch Thousands of Satellites to Provide Internet
around the World." *The Verge*, April 4. https://www.theverge.com/2019/4/4/18295310
/amazon-project-kuiper-satellite-internet-low-earth-orbit-facebook-spacex-starlink.

Precision Agriculture Connectivity Act of 2018. 115th Cong. Pub. L. 115-334.
https://www.congress.gov/bill/115th-congress/house-bill/4881.

Price, Mark, and Stephen Herzenberg. 2011. "The Benefits of State Prevailing Wage Laws." Harrisburg, PA: Keystone Research Center. http://keystoneresearch.org/sites /keystoneresearch.org/files/KRC-PA-Prevailing-Wage-Oct-2011_FINAL.pdf.

Price, Monroe. 2012. "Narratives of Legitimacy." *Trípodos. Facultat de Comunicació i Relacions Internacionals Blanquerna-URL* 1, no. 30: 9–28.

Quaintance, Zack. 2020. "How San Rafael, Calif. Built a Wi-Fi Network during a Pandemic." *Government Technology*, June 17. https://www.govtech.com/network/How -San-Rafael-Calif-Built-a-Wi-Fi-Network-During-a-Pandemic.html.

Raboy, Marc, and David Taras. 2004. "The Politics of Neglect of Canadian Broadcasting Policy." *Policy Options*, March 1. https://policyoptions.irpp.org/magazines /realignment-on-the-right/the-politics-of-neglect-of-canadian-broadcasting-policy.

Raymond, Mark, and Gordon Smith. 2014. "Reimagining the Internet: The Need for High-level Strategic Vision for Internet Governance." In *Organized Chaos: Reimagining the Internet*, edited by Mark Raymond and Gordon Smith, 9–30. Waterloo, Ontario: Center for International Governance Innovation.

Reardon, Marguerite. 2020. "Coronavirus Transforms Peak Internet Usage into the New Normal." *CNET*, March 23. https://www.cnet.com/news/coronavirus-has-made -peak-internet-usage-into-the-new-normal/.

Regional Rural Development Centers and the National Digital Education Extension Team. 2018. "RE: Broadband e-Connectivity Pilot Program." Regulations.gov. https:// beta.regulations.gov/document/RUS-18-TELECOM-0004-0269.

Reglitz, Merten. 2019. "The Human Right to Free Internet Access." *Journal of Applied Philosophy* 37, no. 2: 314–331. https://doi.org/10.1111/japp.12395.

Reibel, Jennifer. 2018. "Manufacturer Consolidation Reshaping the Farm Equipment Marketplace." *Farm Equipment*, August 29. https://www.farm-equipment.com/articles /15962-manufacturer-consolidation-reshaping-the-farm-equipment-marketplace.

Reisinger, Don. 2020. "How 5G Promises to Revolutionize Farming." *Fortune*, February 28. https://fortune.com/2020/02/28/5g-farming.

Rhinesmith, Colin. 2016. *Digital Inclusion and Meaningful Broadband Adoption Initiatives*. Evanston, IL: Benton Institute for Broadband and Society. https://www.benton .org/sites/default/files/broadbandinclusion.pdf.

Rizzato, Francesco, and Ian Fogg. 2020. "How AT&T, Sprint, T-Mobile and Verizon Differ in Their Early 5G Approach." Opensignal. February 20. https://www.opensignal.com /2020/02/20/how-att-sprint-t-mobile-and-verizon-differ-in-their-early-5g-approach.

Roberts, Elizabeth, Brett A. Anderson, Sarah Skerratt, and John Farrington. 2017. "A Review of the Rural-Digital Policy Agenda from a Community Resilience Perspective." *Journal of Rural Studies* 54: 372–385. https://doi.org/10.1016/j.jrurstud.2016.03.001.

Romm, Tony. 2015. "Wired to Fail." *Politico*, July 28. http://politi.co/1fCIzRR.

Rosenblum, Cassady. 2017. "Hillbillies Who Code: The Former Miners Out to Put Kentucky on the Tech Map." *Guardian*, April 21. https://www.theguardian.com/us -news/2017/apr/21/tech-industry-coding-kentucky-hillbillies.

Rosenworcel, Jessica. 2015. "Bridging the Homework Gap." *Huffington Education*, June 15. http://transition.fcc.gov/files/documents/Bridging-the-Homework-Gap-Rosen worcel-Editorial.pdf.

Rosenworcel, Jessica. 2020. "Commissioner Rosenworcel Statement on Rural Digital Opportunity Fund." Federal Communications Commission. January 30. https://www.fcc .gov/document/commissioner-rosenworcel-statement-rural-digital-opportunity-fund.

Rotz, Sarah, Evan Gravely, Ian Mosby, Emily Duncan, Elizabeth Finnis, Mervyn Horgan, Joseph LeBlanc, Ralph Martin, Hannah T. Neufeld, and Andrew Nixon. 2019. "Automated Pastures and the Digital Divide: How Agricultural Technologies Are Shaping Labour and Rural Communities." *Journal of Rural Studies* 68: 112–122. https://doi .org/10.1016/j.jrurstud.2019.01.023.

Rudolph, Richard, and Scott Ridley. 1986. *Power Struggle: The Hundred-year War Over Electricity*. New York: Harper and Row.

Ruiz, Rafico. 2014. "Arctic Infrastructures: Tele Field Notes." *Communication +1* 3, no. 1. https://doi.org/10.7275/R5D21VHD.

Rural Broadband Policy Group. 2009. "Comments of the Rural Broadband Policy Group" (GN Docket No. 09-561). Federal Communications Commission. https:// ecfsapi.fcc.gov/file/6520203513.pdf.

Rural Electrification Administration. 1966. *Rural Lines, USA: The Story of Cooperative Rural Electrification*. Washington, DC: Rural Electrification Administration.

Rural Utilities Service. 1996. "Comments of the Rural Utilities Service" (CC Docket No. 96-45). Federal Communications Commission. https://ecfsapi.fcc.gov/file/1586310001 .pdf.

Rural Utilities Service. 1999. "Reply Comments of the Rural Utilities Service" (CC Docket No. 96-45). Federal Communications Commission. https://ecfsapi.fcc.gov /file/6006041491.pdf.

Rural Utilities Service. 2000a. "Construction and Installation of Broadband Telecommunication Services in Rural America; Availability of Loan Funds." *Federal Register* 65, no. 15: 75920–75921.

Rural Utilities Service. 2000b. "Exparte Comments of the Rural Utilities Service" (CC Docket No. 96-45). Federal Communications Commission. https://ecfsapi.fcc.gov /file/6011255967.pdf.

Rural Utilities Service. 2000c. "Ex Parte Comments of the Rural Utilities Service" (CC Docket No. 96-262, DA No. 00-1882). Federal Communications Commission. https://ecfsapi.fcc.gov/file/6511960044.pdf.

Rural Utilities Service. 2002. "*Ex Parte* Comments of the Rural Utilities Service" (CC Docket No. 96-45). Federal Communications Commission. https://ecfsapi.fcc.gov/file/6513194817.pdf.

Rural Utilities Service. 2012. "Re: GN Docket No. 11-121, Inquiry concerning the Deployment of Advanced Telecommunications Capability to All Americans in a Reasonable and Timely Fashion, and Possible Steps to Accelerate Such Deployment Pursuant to Section 706 of the Telecommunications Act of 1996, as Amended by the Broadband Data Improvement Act." Federal Communications Commission. https://ecfsapi.fcc.gov/file/7021989631.pdf.

Rysavy, Peter. 2018. "How 5G Will Solve Rural Broadband." *FierceWireless*, January 29. https://www.fiercewireless.com/wireless/industry-voices-rysavy-how-5g-will-solve-rural-broadband.

Sadowski, Jathan. 2020. "The Internet of Landlords: Digital Platforms and New Mechanisms of Rentier Capitalism." *Antipode* 52, no. 2: 562–580. https://doi.org/10.1111/anti.12595.

Sallet, Jonathan. 2019a. "Bringing High-Performance Broadband to Rural America." Benton Institute for Broadband and Society. November 13. https://www.benton.org/blog/bringing-high-performance-broadband-rural-america.

Sallet, Jonathan. 2019b. *Broadband for America's Future*. Evanston, IL: Benton Institute for Broadband and Society. https://www.benton.org/sites/default/files/BBA_full_F5_10.30.pdf.

Sallet, Jonathan. 2020. "From Networks to People." Benton Institute for Broadband and Society. January 7. https://www.benton.org/blog/networks-people.

Salter, Jim. 2020. "5G in Rural Areas Bridges a Gap That 4G Doesn't, Especially Low- and Mid-Band." *Ars Technica*, September 14. https://arstechnica.com/features/2020/09/5g-03-rural/.

Sarikakis, Katherine, and Leslie Regan Shade, eds. 2007. *Feminist Interventions in International Communication*. Lanham, MD: Roman & Littlefield.

Satterwhite, Ellen. 2015. "Study Shows Home Values up 3.1% with Access to Fiber." Fiber Broadband Association. June 29. https://www.fiberbroadband.org/blog/study-shows-home-values-up-3.1-with-access-to-fiber.

Schimmelpfennig, David. 2016. *Farm Profits and Adoption of Precision Agriculture* (Economic Research Report No. 217). Washington, DC: US Department of Agriculture. https://www.ers.usda.gov/webdocs/publications/80326/err-217.pdf.

Schneider, Nathan. 2018. *Everything for Everyone: The Radical Tradition That Is Shaping the Next Economy*. New York: Bold Type Books.

Seifer, Angela, and Bill Callahan. 2020. "Limiting Broadband Investment to 'Rural Only' Discriminates against Black Americans and Other Communities of Color." June. National Digital Inclusion Alliance. https://www.digitalinclusion.org/digital-divide-and-systemic-racism/.

Shannon, Kent, David Clay, and Newell Kitchen, eds. 2018. *Precision Agriculture Basics*. Madison, WI: American Society of Agronomy.

Sharpton, Al, Geoffrey Starks, Vanita Gupta, Marc Morial and Maurita Coley. 2020. "Broadband Access Is a Civil Right We Can't Afford to Lose—but Many Can't Afford to Have." *Essence*, June 17. https://www.essence.com/news/broadband-access-is-a-civil-right-we-cant-afford-to-lose-but-many-cant-afford-to-have/.

Smith, Brad. 2018. "The Rural Broadband Divide: An Urgent National Problem That We Can Solve." *Microsoft on the Issues*, December 4. https://blogs.microsoft.com/on-the-issues/2018/12/03/the-rural-broadband-divide-an-urgent-national-problem-that-we-can-solve.

Sohn, Gigi. 2018. "A Policy Framework for an Open Internet Ecosystem." *Georgetown Law Technology Review* 2: 335–359. https://georgetownlawtechreview.org/wp-content/uploads/2018/07/2.2-Sohn-pp-335-59.pdf.

Sohn, Gigi. 2020. "During the Pandemic, the FCC Must Provide Internet for All." *Wired*, April 28. https://www.wired.com/story/opinion-during-covid-19-the-fcc-needs-to-provide-internet-for-all/.

Souter, David, and Anri van der Spuy. 2019. *UNESCO's Internet Universality Indicators: A Framework for Assessing Internet Development*. Paris: UNESCO. https://en.unesco.org/sites/default/files/internet_universality_indicators_print.pdf.

Specht, Doug, and Anna Feigenbaum. 2019. "From the Cartographic Gaze to Contestatory Cartographies." In *Mapping and Politics in the Digital Age*, edited by Pol Bargues-Pedreny, David Chandler, and Elena Simon, 39–55. New York: Routledge.

Spencer, R. 2018. "5G and Smart Farming IoT: Promise of Making the World Green Again. Lanner. July 30. https://www.lanner-america.com/blog/smart-farming-iot-5g-agriculture.

Starosielski, Nicole. 2015. *The Undersea Network*. Durham, NC: Duke University Press.

Stegeman, Jonathan. 2019. *Broadband Mapping Initiative: Proof of Concept*. Seattle, WA: CostQuest Associates. https://www.ustelecom.org/wp-content/uploads/2019/08/USTelecom-Mapping-Pilot-Filing-and-Findings.pdf.

Stella, Shiva. 2019. "Public Knowledge Finds FCC Rural Digital Opportunity Fund Insufficient to End Digital Divide." Public Knowledge. August 1. https://www.publicknowledge

.org/press-release/public-knowledge-finds-fcc-rural-digital-opportunity-fund-insufficient-to-end-digital-divide.

Stenberg, Peter, Mitch Morehart, Stephen Vogel, John Cromartie, Vince Breneman, and Dennis Brown. 2010. *Broadband Internet's Value for Rural America* (Economic Research Report No. 78). Washington, DC: US Department of Agriculture. https://www.ers.usda.gov/webdocs/publications/46200/9335_err78_1_.pdf?v=3513.1.

Stiglitz, Joseph E. 1989. "Markets, Market Failures, and Development." *American Economic Review* 79, no. 2: 197–203.

Stiglitz, Joseph, David Moss, and John Cisternino, 2009. "Regulation and Failure." In *New Perspectives on Regulation,* edited by David Moss and John Cisternino, 13–26. Cambridge, MA: Tobin Project.

Strover, Sharon. 2018. "Reaching Rural America with Broadband Internet Service. *The Conversation,* January 16. http://theconversation.com/reaching-rural-america-with-broadband-internet-service-82488.

Strover, Sharon. 2019. "Public Libraries and 21st Century Digital Equity Goals." *Communication Research and Practice* 5, no. 2: 188–205. https://doi.org/10.1080/22041451.2019.1601487.

Suchman, Mark C. 1995. "Managing Legitimacy: Strategic and Institutional Approaches." *Academy of Management Review* 20, no. 3: 571–610. https://doi.org/10.2307/258788.

Sylvain, Olivier. 2012. "Broadband Localism." *Ohio State Law Journal* 73, no. 4: 795–838.

Taglang, Kevin. 2019. "CenturyLink and Frontier Miss FCC Connect America Fund Broadband Deployment Milestones." Benton Institute for Broadband and Society. January 29. https://www.benton.org/headlines/centurylink-and-frontier-miss-fcc-connect-america-fund-broadband-deployment-milestones.

Tarnoff, Ben. 2016. "The Internet Should Be a Public Good." *Jacobin,* August 31. https://jacobinmag.com/2016/08/internet-public-dns-privatization-icann-netflix.

Taylor, Gregory. 2018. "Remote Rural Broadband Systems in Canada." *Telecommunications Policy* 42, no. 9: 744–756. https://doi.org/10.1016/j.telpol.2018.02.001.

Taylor, Gregory, and Catherine Middleton, eds. 2020. *Frequencies: International Spectrum Policy.* Montreal: McGill-Queens University Press.

Telecommunications Act of 1996. 104th Cong. P.L. 104-104. https://transition.fcc.gov/Reports/tcom1996.txt.

Thomas, Alexander R., Brian Lowe, Gregory Fulkerson, and Polly Smith. 2013. *Critical Rural Theory: Structure, Space, Culture.* Lanham, MD: Lexington Books.

Thompson, Larry, and Warren Vande Stadt. 2017. "5G Is Not the Answer for Rural Broadband." *Broadband Communities*, March–April. https://www.bbcmag.com/rural -broadband/5g-is-not-the-answer-for-rural-broadband.

Tibken, Shara. 2018. "Life in the Slow Lane: Welcome to the Internet in Rural America." *CNET*, October 26. https://www.cnet.com/news/life-in-the-slow-lane-welcome -to-the-internet-in-rural-america.

Tita, Bob. 2019. "Deere Turns to U.S. after Growth Stalls Overseas." *Wall Street Journal*, September 22. https://www.wsj.com/articles/deere-turns-to-u-s-after-growth-stalls -overseas-115691.50000.

Tobey, Ronald. C. 1996. *Technology as Freedom: The New Deal and the Electrical Modernization of the American Home*. Berkeley: University of California Press.

Tonsager, Dallas. 2012. "Letter to the Government Accountability Office re: Report 12–937." In *Recovery Act: Broadband Programs Are Ongoing, and Agencies' Efforts Would Benefit from Improved Data Quality* (GAO-12-937) by the Government Accountability Office, 32–37. Washington, DC: Government Accountability Office. http://www.gao .gov/products/GAO-12-937.

Townsend, Leanne, Arjuna Sathiaseelan, Gorry Fairhurst, and Claire Wallace. 2013. "Enhanced Broadband Access as a Solution to the Social and Economic Problems of the Rural Digital Divide." *Local Economy* 28, no. 6: 580–595. https://doi.org/10.1177 /0269094213496974.

Townsend, Leanne, Claire Wallace, and Gorry Fairhurst. 2015. "'Stuck Out Here': The Critical Role of Broadband for Remote Rural Places." *Scottish Geographical Journal* 131, nos. 3–4: 171–180. https://doi.org/10.1080/14702541.2014.978807.

Treacy, Ann. 2017. "Broadband Is a Public Good." *Blandin on Broadband*, January 30. https://blandinonbroadband.org/2017/01/30/broadband-is-a-public-good-government -support-makes-sense/.

Treacy, Ann. 2018a. "Despite What the Telecom Industry Says, RS Fiber Is a House of Opportunity, Not a House of Cards." *Blandin on Broadband*, September 3. https:// blandinonbroadband.org/2018/09/03/despite-what-the-telecom-industry-says-rs -fiber-is-a-house-of-opportunity-not-a-house-of-cards.

Treacy, Ann. 2018b. "MN Telecom Alliance Response to RS Fiber Financial News." *Blandin on Broadband*, August 24. https://blandinonbroadband.org/2018/08/24/mn -telecom-alliance-response-to-rs-fiber-financial-news.

Treacy, Ann. 2019. "CenturyLink Misses FCC Connect America Fund Milestones in Minnesota. *Blandin on Broadband*, January 31. https://blandinonbroadband.org/2019 /01/31/centurylink-misses-fcc-connect-america-fund-milestones-in-minnesota.

Trostle, H., Katie Kienbaum, Michelle Andrews, Ny Ony Razafindrable, and Christopher Mitchell. 2020. *Cooperatives Fiberize Rural America: A Trusted Model for the*

Internet Era. Washington, DC: Institute for Local Self-Reliance. https://cdn.ilsr.org
/wp-content/uploads/2020/05/2020_05_19_Rural-Co-op-Report.pdf.

Trump, Donald. 2018. "Presidential Executive Order on Streamlining and Expedit-
ing Requests to Locate Broadband Facilities in Rural America." Whitehouse.gov.
January 8. https://www.whitehouse.gov/presidential-actions/presidential-executive
-order-streamlining-expediting-requests-locate-broadband-facilities-rural-america.

Tucker, Russell, Joseph Goodenbery, and Katherine Loving. 2018. *Unlocking the Value
of Broadband for Electric Cooperative Consumer-Members.* Arlington, VA: National Rural
Electric Cooperative Association. http://www.electric.coop/wp-content/uploads/2018
/09/Unlocking-the-Value-of-Broadband-for-Co-op-Consumer-Members_Sept_2018
.pdf.

Turner, S. Derek. 2016. *Digital Denied: The Impact of Systemic Racial Discrimination on
Home-Internet Adoption.* New York: Free Press. https://www.freepress.net/sites/default
/files/legacy-policy/digital_denied_free_press_report_december_2016.pdf.

Turow, Joseph. 2013. *The Daily You: How the New Advertising Industry Is Defining Your
Identity and Your Worth.* New Haven, CT: Yale University Press.

Tzach, Tomer. 2018. "Soil Sensors: A New Direction in Precision Agriculture to
Improve Crop Production." *PrecisionAg,* April 10. https://www.precisionag.com/in-field
-technologies/connectivity/soil-sensors-a-new-direction-in-precision-agriculture-to
-improve-crop-production.

United Nations. 2016. "The Promotion, Protection and Enjoyment of Human
Rights on the Internet." Human Rights Council. https://www.article19.org/data/files
/Internet_Statement_Adopted.pdf.

United Soybean Board. 2019. *Rural Broadband and the American Farmer.* Chesterfield,
MO: United Soybean Board. https://api.unitedsoybean.org/uploads/documents
/58546-1-ruralbroadband-whitepages-final.pdf.

*United States of America v. Deere & Company, Precision Planting LLC and Monsanto
Company.* Complaint (N.D. Ill. 2016). https://www.justice.gov/atr/case-document
/file/905571/download.

Universal Service Administrative Company. 2019. *2019 Annual Report.* Washington,
DC: Universal Service Administrative Company. https://www.usac.org/wp-content
/uploads/about/documents/annual-reports/2019/USAC-2019-Annual-Report.pdf.

UScellular. 2019. "Reply Comments of United States Cellular Corporation" (WC
Docket Nos. 19-126, 10-90). Federal Communications Commission. https://ecfsapi
.fcc.gov/file/1021019362798/2019%201021%20USCC%20RDOF%20Reply%20-%20
FINAL%20As%20Filed.pdf.

US Census Bureau. n.d. "Population: 1790 to 1990." https://www.census.gov/popula
tion/www/censusdata/files/table-4.pdf.

US Census Bureau. 2017. "What Is Rural America?" https://www.census.gov/library /stories/2017/08/rural-america.html.

US Congress. House of Representatives. Committee on Agriculture. 2015. *Big Data and Agriculture: Innovation and Implications.* 114th Cong., 1st sess., October 28. https:// republicans-agriculture.house.gov/uploadedfiles/10.28.15_hearing_transcript.pdf.

US Congress. Senate. Committee on Commerce, Science, and Transportation. 2009. *Oversight of the Broadband Stimulus Programs in the American Recovery and Reinvestment Act.* 111th Cong., 1st sess., October 27. https://www.govinfo.gov/content/pkg /CHRG-111shrg55984/html/CHRG-111shrg55984.htm.

US Congress. Senate. Subcommittee on Communications, Technology, Innovation and the Internet. 2016. *Ensuring Intermodal USF Support for Rural America.* 114th Cong., 2nd sess., February 4. https://www.govinfo.gov/content/pkg/CHRG-114shrg24144/html /CHRG-114shrg24144.htm.

US Department of Agriculture. n.d. "Eligible Service Area." https://www.usda.gov /reconnect/eligible-service-area.

US Department of Agriculture. 2005. *Audit Report: Rural Utilities Service Broadband Grant and Loan Programs* (No. 09601-4-Te). Washington, DC: US Department of Agriculture. https://corpora.tika.apache.org/base/docs/govdocs1/153/153075.pdf.

US Department of Agriculture. 2013. *Report on the Definition of "Rural."* Washington, DC: US Department of Agriculture. https://www.rd.usda.gov/files/RDRural Definition-ReportFeb2013.pdf.

US Department of Agriculture. 2017. *Rural America at a Glance: 2017 Edition* (Economic Information Bulletin 182). Washington, DC: US Department of Agriculture. https://www.ers.usda.gov/webdocs/publications/85740/eib-182.pdf.

US Department of Agriculture. 2018a. *2019 Budget Summary.* Washington, DC: US Department of Agriculture. https://www.usda.gov/sites/default/files/documents /usda-fy19-budget-summary.pdf.

US Department of Agriculture. 2018b. *Rural Broadband Access Loan Application Guide Fiscal Year 2018.* Washington, DC: US Department of Agriculture. https://www.rd .usda.gov/files/FB_AppGuide_Revised_032518_1.pdf.

US Department of Agriculture. 2019a. *A Case for Rural Broadband: Insights on Rural Broadband Infrastructure and Next Generation Precision Agriculture Technologies.* Washington, DC: US Department of Agriculture. https://www.usda.gov/sites/default/files /documents/case-for-rural-broadband.pdf.

US Department of Agriculture. 2019b. *Farm Computer Usage and Ownership.* Washington, DC: US Department of Agriculture. https://downloads.usda.library.cornell.edu /usda-esmis/files/h128nd689/8910k592p/qz20t442b/fmpc0819.pdf.

US Department of Agriculture. 2020. "ReConnect Program Awardees." https://www.usda.gov/reconnect/awardees.

USTelecom. 2018. "Comments of USTelecom—The Broadband Association" (Docket No. RUS-18-TELECOM-0004). https://www.ustelecom.org/wp-content/uploads/docu ments/9-10-18%20USTelecom%20RUS%20Pilot%20Comments.pdf.

Vaidhyanathan, Siva. 2018. *Antisocial Media: How Facebook Disconnects Us and Undermines Democracy*. Oxford: Oxford University Press.

Van den Bulck, Hilde. 2012. "Stakeholder Analysis in Media Studies." In *Trends in Communication Policy Research: New Theories, Methods and Subjects*, edited by Natascha Just and Manuel Puppis, 217–232. Bristol, UK: Intellect.

Van den Bulck, Hilde, and Karen Donders. 2014. "Of Discourses, Stakeholders and Advocacy Coalitions in Media Policy: Tracing Negotiations towards the New Management Contract of Flemish Public Broadcaster VRT." *European Journal of Communication* 29, no. 1: 83–99. https://doi.org/10.1177/0267323113509362.

van Dijk, Jan. 2005. *The Deepening Divide: Inequality in the Information Society*. Thousand Oaks, CA: SAGE.

van Dijk, Jan. 2020. *The Digital Divide*. Cambridge: Polity.

van Dijk, Teun. 1993. "Principles of Critical Discourse Analysis." *Discourse & Society* 4, no. 2: 249–283. https://doi.org/10.1177/0957926593004002006.

van Shewick, Barbara. 2007. "Towards an Economic Framework for Network Neutrality Regulation." *Journal of Telecommunications and High Technology Law* 5, no. 2: 329–391. https://cyberlaw.stanford.edu/files/publication/files/vanschewick-2007-towards-an -economic-framework.pdf.

Viasat. 2019. "Notice of *Ex Parte* Presentation of Viasat Inc."). National Exchange Carrier Association. December 20 https://prodnet.www.neca.org/publicationsdocs /wwpdf/122319viasat.pdf.

Voltaire. 1772. *La Begueule*. https://fr.wikisource.org/wiki/Contes_en_vers_(Voltaire) /La_Bégueule.

Wanstreet, Rian. 2018. "America's Farmers Are Becoming Prisoners to Agriculture's Technological Revolution." *Vice*, March 8. https://www.vice.com/en_us/article/a34 pp4/john-deere-tractor-hacking-big-data-surveillance.

Warren, Elizabeth. 2019. "My Plan to Invest in Rural America." *Medium*, August 7. https://medium.com/@teamwarren/my-plan-to-invest-in-rural-america-94e3a80 d88aa.

Wasko, Janet, Graham Murdock, and Helena Sousa. 2011. *The Handbook of Political Economy of Communications*. Chichester, UK: Wiley-Blackwell.

Whitacre, Brian. 2016. "Technology Is Improving—Why Is Rural Broadband Access Still a Problem? *The Conversation*, June 8. https://theconversation.com/technology-is -improving-why-is-rural-broadband-access-still-a-problem-60423.

Whitacre, Brian. 2017. "Fixed Broadband or Mobile: What Makes Us More Civically Engaged?" *Telematics and Informatics* 34, no. 5: 755–766.

Whitacre, Brian, and Roberto Gallardo. 2020. "State Broadband Policy: Impacts on Availability." *Telecommunications Policy* 44, no. 9. https://doi.org/10.1016/j.telpol.2020 .102025.

Whitacre, Brian, Roberto Gallardo, Angela Siefer, and Bill Callahan. 2018. "The FCC's Blurry Vision of Satellite Broadband." *Daily Yonder*, March 26. https://www .dailyyonder.com/fccs-blurry-vision-satellite-broadband/2018/03/26/24739.

Whitacre, Brian, Roberto Gallardo, and Sharon Strover. 2014. "Broadband's Contribution to Economic Growth in Rural Areas: Moving towards a Causal Relationship." *Telecommunications Policy* 38, no. 11: 1011–1023. https://doi.org/10.1016/j.telpol .2014.05.005.

Whitacre, Brian. E., and Phumsith Mahasuweerachai. 2008. "'Small' Broadband Providers and Federal Assistance Programs: Solving the Digital Divide?" *Journal of Regional Analysis and Policy* 38, no. 3: 1–15.

Whitacre, Brian. E., and Jacob L. Manlove. 2016. "Broadband and Civic Engagement in Rural Areas: What Matters? *Community Development* 47, no. 5: 700–717. https:// doi.org/10.1080/15575330.2016.1212910.

Whitacre, Brian E., Tyler B. Mark, and Terry W. Griffin. 2014. "How Connected Are Our Farms?" *Choices* 29, no. 3: 1–9.

Whitacre, Brian E., Denna Wheeler, and Chad Landgraf. 2016. "What Can the National Broadband Map Tell Us about the Health Care Connectivity Gap." *Journal of Rural Health* 33, no. 3: 284–289. https://doi.org/10.1111/jrh.12177.

Wicker, Robert, Joe Manchin, Tammy Baldwin, Roy Blunt, Richard Burr, Shelley Moore Capito, Steve Daines, et al. 2016. "To Honorable Tom Wheeler, Chairman of the Federal Communications Commission." July 11 Washington, DC: US Senate. https://www.manchin.senate.gov/imo/media/doc/071116%20Mobility%20 Fund%20-%20Ag%20Letter.pdf?cb.

Wiens, Kyle. 2015a. "New High-Tech Farm Equipment Is a Nightmare for Farmers. *Wired*, February 5. https://www.wired.com/2015/02/new-high-tech-farm-equipment -nightmare-farmers.

Wiens, Kyle. 2015b. "We Can't Let John Deere Destroy the Very Idea of Ownership." *Wired*, April 21. https://www.wired.com/2015/04/dmca-ownership-john-deere.

Wiens, Kyle, and Elizabeth Chamberlain. 2018. "John Deere Just Cost Farmers Their Right to Repair." *Wired*, September 19. https://www.wired.com/story/john-deere -farmers-right-to-repair.

Wikipedia. 2020. "Luverne, Minnesota." Last modified August 19, 2020. https://en .wikipedia.org/wiki/Luverne,_Minnesota.

WildBlue Communications and Intelsat Corporation. 2009. "Comments of Wild-Blue Communications, Inc. and Intelsat Corporation" (GN Docket No. 09-29). Federal Communications Commission. https://ecfsapi.fcc.gov/file/6520203432.pdf.

Wilken, Rowan. 2011. *Teletechnologies, Place, and Community*. New York: Routledge.

Williams, Raymond. 1975. *The Country and the City*. Oxford: Oxford University Press.

Wilson, Matthew W. 2017. *New Lines: Critical GIS and the Trouble of the Map*. Minneapolis: University of Minnesota Press.

Winseck, Dwayne, and Dal Young Jin, eds. 2011. *The Political Economies of Media: The Transformation of the Global Media Industries*. New York: Bloomsbury.

Wiseman, Leanne, Jay Sanderson, Airong Zhang, and Emma Jakku. 2019. "Farmers and Their Data: An Examination of Farmers' Reluctance to Share Their Data through the Lens of the Laws Impacting Smart Farming." *NJAS: Wageningen Journal of Life Sciences* 90–91. https://doi.org/10.1016/j.njas.2019.04.007.

W. K. Kellogg Foundation. 2001. "Perceptions of Rural America." December 9. https:// www.wkkf.org:443/resource-directory/resource/2002/12/perceptions-of-rural-america.

Wolfert, Sjaak, Lan Ge, Cor Verdouw, and Marc-Jeroen Bogaardt. 2017. "Big Data in Smart Farming: A Review." *Agricultural Systems* 153: 69–80. https://doi.org/10.1016/j .agsy.2017.01.023.

Woods, Michael. 2005. *Rural Geography*. Thousand Oaks, CA: SAGE.

Wu, Timothy. 2010. *The Master Switch: The Rise and Fall of Information Empires*. New York: Alfred A. Knopf.

Wunsch, Silke. 2013. "Internet Access Declared a Basic Right in Germany." *Deutsche Welle*, January 27. https://www.dw.com/en/internet-access-declared-a-basic-right-in -germany/a-16553916.

Yoo, Christopher. 2006. "Network Neutrality and the Economics of Congestion." *Georgetown Law Journal* 94, no. 6: 1847–1908.

Yoo, Christopher, and Timothy Pfenninger. 2017a. *Municipal Fiber in the United States: An Empirical Assessment of Financial Performance*. Philadelphia: University of Pennsylvania Law School, Center for Technology, Innovation and Competition. https://www .law.upenn.edu/live/files/6611-report-municipal-fiber-in-the-united-states-an.

Yoo, Christopher, and Timothy Pfenninger. 2017b. "Correction." Philadelphia: University of Pennsylvania Law School, May 26. http://www.baller.com/wp-content /uploads/Press-Release_Penn-Law_5-26-17.pdf.

Yoo, Christopher, and Timothy Pfenninger. 2017c. "Municipal Fiber in the United States: Response to Critics and Extension of the Analysis." Philadelphia: University of Pennsylvania Law School, Center for Technology, Innovation and Competition. https://www.law.upenn.edu/live/files/6674-yoo-pfenninger---response.

Zimmer, Jameson. 2018. "FCC Broadband Definition Has Changed before and Will Change Again." BroadbandNow. February 18. https://broadbandnow.com/report /fcc-broadband-definition/.

Index

Information Policy Series

Edited by Sandra Braman

T Press
net Rossi
5 Main Street, 9th floor
A, 02142

T.edu
1ett@mit.edu
7-253-2882

e authorized representative in the EU for product safety and compliance is

sy Access System Europe Oü, 16879218
ustamäe tee 50,
Z, 10621

sr.requests@easproject.com
72 56 968 939

3N: 9780262543064
lease ID: 152941459

www.ingramcontent.com/pod-product-compliance
Lightning Source LLC
Chambersburg PA
CBHW022302280326
41932CB00010B/954